AN AGRICULTURAL GEOGRAPHY OF GREAT BRITAIN

BELL'S ADVANCED ECONOMIC GEOGRAPHIES

General Editor
PROFESSOR R. O. BUCHANAN
M.A.(N.Z.), B.Sc.(Econ.), Ph.D.(London)
Professor Emeritus, University of London

A. Systematic Studies

AN ECONOMIC GEOGRAPHY OF OIL
Peter R. Odell, B.A., Ph.D.

PLANTATION AGRICULTURE
P. P. Courtenay, B.A., Ph.D.

NEW ENGLAND: A STUDY IN INDUSTRIAL ADJUSTMENT
R. C. Estall, B.Sc.(Econ.), Ph.D.

GREATER LONDON: AN INDUSTRIAL GEOGRAPHY
J. E. Martin, B.Sc.(Econ.), Ph.D.

GEOGRAPHY AND ECONOMICS
Michael Chisholm, M.A.

AGRICULTURAL GEOGRAPHY
Leslie Symons, B.Sc.(Econ.), Ph.D.

REGIONAL ANALYSIS AND ECONOMIC GEOGRAPHY
John N. H. Britton, M.A., Ph.D.

B. Regional Studies

AN ECONOMIC GEOGRAPHY OF EAST AFRICA
A. M. O'Connor, B.A., Ph.D.

AN ECONOMIC GEOGRAPHY OF WEST AFRICA
H. P. White, M.A., and M. B. Gleave, M.A.

YUGOSLAVIA: PATTERNS OF ECONOMIC ACTIVITY
F. E. Ian Hamilton, B.Sc.(Econ.), Ph.D.

AN AGRICULTURAL GEOGRAPHY OF GREAT BRITAIN
J. T. Coppock, M.A., Ph.D.

AN HISTORICAL INTRODUCTION TO THE ECONOMIC GEOGRAPHY OF GREAT BRITAIN
Wilfred Smith, M.A.

THE BRITISH IRON & STEEL SHEET INDUSTRY SINCE 1840
Kenneth Warren, M.A., Ph.D.

C630.912/KOP

An Agricultural Geography of Great Britain

J. T. COPPOCK

M.A., Ph.D.

Ogilvie Professor of Geography
University of Edinburgh

LONDON

G. BELL & SONS, LTD

1971

Copyright © 1971 by
G. BELL AND SONS, LTD
York House, Portugal Street
London, W.C.2

All Rights Reserved. No part of this publication may be reproduced, stored in a retrieval system, or transmitted, in any form or by any means, electronic, mechanical, photocopying, recording or otherwise, without the prior permission of G. Bell & Sons Ltd.

ISBN 0 7135 1950 9

Printed in Great Britain by
NEILL & CO. LTD., EDINBURGH.

Preface

I am grateful to Professor R. O. Buchanan for inviting me to write this book because it has forced me to look afresh at what I have been teaching in the field of agricultural geography. The lecture is regarded by some dons as the ultimate goal of all academic efforts, in theory if not in practice; yet, whatever the truth of the dictum 'It's not what you say, but the way that you say it', it is often all too easy to deceive oneself by words which, however fine they sound in the lecture theatre, simply will not do in cold print and fortunately either vanish into thin air as soon as they are spoken or are retained only in garbled form in student notebooks. Although I am critical of the excesses of the 'publish or perish' school, there is obvious merit and discipline in trying to express thoughts and material in more permanent form. There is also social gain, and my task in writing this book would have been far easier if much more of the effort and research which has been devoted to the preparation of courses in agricultural geography in British universities over the past twenty years had been committed to print. As it is, the literature on the agricultural geography of Great Britain is, with the exception of the work of the Land Utilisation Survey, extraordinarily sparse, a fact of which I have become vividly aware in writing this book.

It is true that there is a vast literature about British agriculture, but most of it is concerned with technical and scientific aspects, often in very narrow and specialised fields; and, although in total it contains much material relevant to an understanding of the agricultural geography of Great Britain, the task of sifting this vast literature is beyond the capacity of any individual. It is also true that an immense amount of statistical information is collected each year by government departments which, if fully analysed, would provide the raw material for a very satisfactory agricultural geography; but, until recently, government departments have not been very interested in the spatial characteristics of British agriculture, and universities have not been geared to undertake the analyses of their data. Fortunately, the advent of the digital computer has transformed this situation and I am hopeful that, with the support of the Social Science Research Council and the co-operation of the Ministry of Agriculture, Fisheries and Food and the Department of Agriculture and Fisheries for Scotland, we shall be able to remedy some of these deficiencies.

This book is concerned with the geography of agriculture in Great Britain. In it I describe and seek to explain (as far as this is possible) the variations in agricultural activity throughout Great Britain. This is, of course, a vast undertaking, as I am often reminded by my agricultural colleagues who explore in depth some aspect of agriculture, and it is not made any easier by the rapidity of change in the post-war period. I have sometimes felt like a marksman trying simultaneously to observe a large number of targets all moving rapidly at different speeds and in different directions. I began to prepare this book late in 1966 and much of the statistical material on which the maps are based relates to 1965; most of the writing was done in 1968 and 1969. Of course, a synoptic view of this kind must inevitably look back to what has recently been. In this sense all geography is historical geography and a truly contemporaneous account is neither technically nor humanly possible. Similarly, projections of trends into the future, hazardous as they are for the whole country, are infinitely more so at the regional and local level. I have tried to steer a course between the ephemeral, which is likely to reflect only the fashion of the present, and the truly historical, which relates to conditions which no longer exist. The perspective of this book is therefore somewhat longer than that of most agricultural writing, but I hope that the discrepancies between what is and what recently has been are not too great.

Many of these difficulties would arise if I were an agriculturalist writing for an agricultural readership; but I am a geographer, writing for geographers. While I hope that this book will be of interest to all those, whether administrators, economists, scientists and farmers, who are concerned with how the land of Britain is used for the production of crops, livestock and livestock products, I must ask their indulgence for a certain technical naïveté. Professor G. P. Wibberley's strictures that 'many geographers have only a superficial acquaintance with the technical and commercial aspects of food production' has much justification, for geographers receive no formal training in agriculture as such. I myself had none, although I did attend courses given by Sir Frank Engledow and his colleagues at the Cambridge School of Agriculture. I have since learnt much from farmers, officials and fellow academics, although none of these is responsible for any misconceptions I have. Given this lack of training, it is difficult to know how much elementary technical explanation to introduce into a book of this kind. In my view, formal instruction in such matters has no place here and those who seek it must look to elementary text books on agriculture and agricultural economics; but I have felt it necessary to make numerous observations which agriculturalists might regard as self-evident. Equally, I must ask the indulgence of my geographical colleagues if I have strayed too far along the technical path; the agricultural literature is so overwhelmingly technical and the geographical literature so sparse, that some tendency in this direction is inevitable. I have tried to avoid the technical for its own sake and to remember that my primary task is one of explanatory description.

If agriculture is changing rapidly, so too is the nature of geography. This new geography takes two related forms. On the one hand, there is the attempt to devise a more explicit theoretical basis for the subject, in which the underlying spatial order of phenomena on the earth's surface can be more easily appreciated, an attempt which has been understandably more successful in the field of urban and transport geography. On the other, there is the development of more precise methods of measuring areal distribution and association in place of what Professor P. Crowe has called the 'eye-ball' method. I am in sympathy with these approaches, provided that the ultimate aim is a better understanding of the real world and not model building for its own sake. In this ideal world (with which geographers are not ultimately concerned) I should like to have written a book which took account of these changes; but the ground work has not yet been done in the field of agricultural geography, the statistical data often do not exist and I have not the technical competence to undertake it.

PREFACE

In writing this book I have trespassed on the time of many busy people and received much helpful advice on matters on which they are expert. I am particularly grateful to those who read and commented on various sections of the manuscript: Mr. G. Allanson; Dr. D. B. Grigg; Dr. J. M. M. Cunningham; Mr. J. Eadie; Dr. R. R. W. Folley; Dr. J. C. Holmes; Mr. J. M. Manson; Dr. J. D. W. McQueen; Mr. R. F. Thow; and especially Professor Sir Stephen Watson, who read the whole book. I have not always been able to accept their advice, partly for reasons of space, and there are regrettably many bald statements that ought to be qualified. I am grateful also to those who supplied me with information, especially to Mr. P. Scola and Dr. J. Dunn and their colleagues of the Department of Agriculture and Fisheries for Scotland and Mr. D. Salton and Mr. P. Horscroft and their colleagues of the Ministry of Agriculture, Fisheries and Food. Lastly, I owe a special debt of gratitude to Mr. J. McG. Hotson and Mr. W. A. Sentance, who wrote the computer programs on which some of the maps are based; to Miss S. Bonus, Miss L. Park, Miss A. Souttar and Miss L. Walker, who helped with much of the tedious calculation and plotting of the maps and to Mr. R. Harris who drew them; to Mrs. N. Ogilvie, Mrs. F. Nelson, Miss V. Eddy and Miss A. Haigh, who all contributed by typing from illegible manuscripts or incomprehensible tapes; to my wife and daughter, who corrected my spelling, my punctuation and the wording of my drafts; and to Professor Buchanan, who wielded his editorial pencil with tact and skill. To all these I express my gratitude; any errors are my responsibility alone.

NOTE

Ministry of Agriculture, Fisheries and Food is generally abbreviated to Ministry of Agriculture, and the Department of Agriculture and Fisheries for Scotland to the Department of Agriculture; the term 'the agricultural departments' is used to describe them both.

Where points of the compass or other terms are used to describe a general locality, they are shown in lower case, e.g. east Scotland; where they describe an officially designated area, with defined boundaries, they are shown in upper case, e.g., South-west England for the Ministry of Agriculture's region comprising the counties of Cornwall, Devon, Dorset, Gloucester, Somerset and Wiltshire. For variety, the term 'Britain' is used occasionally as a synonym for 'Great Britain'.

Percentages in tables do not always total 100 per cent. because of rounding.

Contents

		Page
	Preface	v
	List of Maps	xii
	List of Tables	xiii
1.	Introduction	1
2.	The Changing Context of British Farming	11
3.	Land and Weather	29
4.	Farms and Fields	54
5.	Men and Machines	80
6.	Markets and Marketing	107
7.	Land and Livestock	135
8.	Dairying	152
9.	Beef Cattle	179
10.	Sheep and Lambs	200
11.	Pigs and Poultry	222
12.	Crops	245
13.	Horticulture	274
14.	The Pattern of Farming	303
	Postscript	325
	Index	331

Maps and Diagrams

		Page
1a.	Indicators of change, 1867–1967	15
1b.	Indicators of change, 1867–1967	19
2.	Climate	35
3.	Land and land quality	42
4.	Lime and fertiliser	48
5.	Holdings and tenure	59
6.	Labour in agriculture	89
7.	Mechanisation in agriculture	100
8.	Markets for agricultural produce	118
9.	Agricultural land use	137
10.	Agricultural changes, 1865–1965	147
11.	Livestock	150
12.	Dairying	154
13.	Beef cattle	187
14.	Sheep	203
15.	Pigs and poultry	226
16.	Crop acreages	250
17.	Crop yields	251
18.	Barley varieties	255
19.	Horticultural crops	281
20.	Major enterprises	308
21.	Enterprises and man-days	311
22.	Types of farming	317

Tables

		Page
1.	Sample harvest dates	36
2.	Land quality in Great Britain	51
3.	Size of holdings in the Regions of England and in Wales, 1966 and 1967	64
4.	Average size of holding and farming type in Scotland, 1962	65
5.	Percentage distribution of enterprises by size of holding in England and Wales, 1966	67
6.	Percentage of land occupied by fields of different sizes	76
7.	Full-time labour in each Region in England and in Wales, 1967	88
8.	Full-time labour in each Region in Scotland, 1967	90
9.	Casual labour in England and Wales, 1967	91
10.	Family labour in England and Wales, 1967	92
11.	Dairy herds by size groups, 1965	155
12.	Breeds of dairy cows, 1965	156
13.	Production and marketing of milk, England and Wales	171
14.	Dairy cows in each Region of England and in Wales, 1966	174
15.	Dairy cows in each Region of Scotland, 1962	175
16.	Beef cows and 'other male cattle' in each Region of England and in Wales, 1966	194
17.	Beef cattle in Scotland, 1967	196
18.	Fat cattle in Scotland, 1963	198
19.	Percentage of holdings with sheep, by Regions	202
20.	Percentage of holdings with sheep, by size of holding	209
21.	Percentage of gross output due to sheep and wool, 1966	219
22.	Percentage of holdings returning breeding ewes, by type of farm, 1966	220
23.	Percentage of breeding sheep in each Region of England and in Wales, 1966	221
24.	Pigs in herds of different sizes in 1967 and 1968	227
25.	Changes in herd size in England and Wales, 1955–66	232
26.	Size of flocks of adult fowl in England and Wales	240
27.	Cash crops and farm type	270

		Page
28.	Wheat, barley and potatoes, and type of farming, England and Wales, 1966	271
29.	Tillage, wheat, barley, potatoes and sugar beet, England and Wales, 1966	272
30.	Sales of farm crops, 1964	272
31.	Distribution of horticultural output in Great Britain, 1965–6	279
32.	Vegetables in the Regions of England and in Wales, 1968	283
33.	Vegetable acreage groups in England and Wales, 1960 and 1968	283
34.	Orchards in England and Wales, 1966	290
35.	Degree of dominance of enterprise on holdings in England and Wales, 1960	312
36.	Numbers of holdings of each type in 1965	313
37.	Percentage distribution of the main enterprises by type of farm, 1965	320
38.	Types of farming in Scotland, 1962	321
39.	Types of farming in England and Wales, 1966	322

CHAPTER 1

Introduction

Farming is obviously not uniform throughout Great Britain and the various types of farms are not randomly distributed, as an hour's journey by air from Edinburgh to Gatwick will confirm. This flight provides a not unrepresentative transect across Great Britain: a fleeting glimpse of the Lothians, with their large farmsteads and extensive tracts of arable land under cereals and vegetables; the Southern Uplands, dotted with sheep and, less frequently, cattle and with green ribbons of improved land along the valleys; the Lake District, more rugged and deeply dissected than the Southern Uplands, with sharper contrasts in agricultural land use; the small-scale landscape of the livestock and dairy farms of north Lancashire and Craven; the southern Pennines, spotted with reservoirs and with little visible sign of any agricultural use, followed rapidly by the limestone walls and grass fields of the Peak District dairy farms; the mixed farming of the red Midlands, the rapidly changing landscapes of the scarplands, from the arable land of the Northampton Heights, across the grass fields of the Vale of Aylesbury, to the mixed farming of the Chilterns; and then, depending on the approach to Gatwick, the glasshouses of the Lea Valley, the marsh pastures of the lower Thames estuary and the patchwork of market garden crops and orchards of north Kent, or the continuous arable of the Hampshire chalklands, brown, green or yellow according to season; and, finally, the mixed farming of the Weald, with a distinctive admixture of horse stables and broiler houses. Except for the Midlands, where the pattern is less clear-cut, it would not be difficult from this and similar journeys to sketch the outlines of broad agricultural regions which contrasted quite markedly with one another. Defining and understanding these differences is the business of the agricultural geographer.

It may be that this broad impression is unduly influenced by land use, that the apparent homogeneity within regions is illusory and that the agricultural differences between them are small. An observer on foot is as likely to notice the differences between farms as the similarity of farming over large areas; indeed, given some 150,000 farmers cultivating several million fields in a wide range of natural and man-made conditions, marked local differences are to be expected. Such variety need not invalidate the wider view, for there may well be major trends which are obscured by background 'noise'; just as close inspection of a newspaper photograph suggests a haphazard scatter of dots, so the difference in impression from air and ground may be due simply to differences in the scale of observation. Nevertheless, impression is not a secure basis for analysing the variety of farming in Great Britain, and one must measure and examine the range of differences from place to place. In doing so, significant similarities and differences in farming may well come to light which are either not readily observable or not immediately apparent in an impressionistic view.

DESCRIPTION & EXPLANATION IN AGRICULTURAL GEOGRAPHY

Recording and measuring such differences are not, of course, the sole or indeed the main aim of the agricultural geographer, but they are necessarily, in John Weaver's words, 'the first order of geographic business'. Here an immediate difficulty arises. Unlike the laboratory scientist (but like the ecologist and the geologist), the geographer investigates phenomena very much larger than himself and, until the advent of the aeroplane and the satellite, he could not directly observe them, let alone study them. Even now the mental picture of the agriculture of any area acquired by first-hand observation must be the cumulative result of a sequence of impressions and observations, made by travel throughout the area and recorded on maps and in notebooks or, less certainly, in memory. How large an area can be known intimately at first hand in this way is debatable. Agricultural advisory officers, land agents of large estates and others who visit farms frequently, may gain a surprisingly detailed knowledge of their areas and be able to identify the characteristics of individual farms and even of individual fields; but it is doubtful whether even they can know an area as large as a single county in this way. For the academic geographer such first-hand knowledge is generally not possible even for a much smaller area; yet without an adequate general view of the area he is studying, he is like the old lady who, when asked why she did not think before she spoke, protested 'How can I know what I think till I hear what I say?'. Since adequate first-hand knowledge of a country the size of Great Britain is impossible, the agricultural geographer must obtain his knowledge of the pattern of agriculture largely from data collected by others, supplemented as far as possible by personal enquiry. Such material will be mainly statistical, for this alone provides reasonably uniform and contemporaneous data about large areas.

This dependence on secondary data creates another difficulty. The operational unit in farming is the individual farm, yet nearly all data are available either by fields or, more commonly, by some aggregation of farms by parishes, counties or other administrative units, so that the geographer is dealing, not with actual farms, but with a 'parish' farm or a 'county' farm, representing the average of all farms in that area. Generally, farm types and enterprises can be identified only by allocating the individual categories of crops and livestock in these administrative areas to particular enterprises somewhat arbitrarily, or by using a crop or a class of livestock to represent the enterprise to which it belongs, e.g. other male cattle of one year old and over as an index of cattle fattening. These limitations obviously affect the quality of the answers to the questions 'Where?' and 'Why there?'

Given the complexity of British farming and the nature of the available data, one must adopt first an analytic approach and consider these questions in respect of individual enterprises or even crops or classes of livestock. A preliminary answer to the question 'Where?' can be given simply by plotting the spatial distribution of each enterprise. If the study is sufficiently detailed, the characteristics of the distribution may be further analysed by measuring its texture and degree of localisation, though such procedures are justifiable only if the available data are of high quality. It also helps to know on what proportion of farms the enterprise occurs and whether this varies systematically from place to place, for there is no necessary connection between frequency of occurrence and degree of localisation; unfortunately, this information, too, is generally available only for very large regions. Further light can be thrown on each distribution by considering it not only absolutely, but also in relation to other enterprises; for this provides some measure of their competitive strength.

Careful analysis of the spatial distribution of enterprises in this way will suggest some possible reasons, but providing satisfactory answers to the question 'Why there?' is much more difficult. Agriculture is a complex of activities in which each crop and class of livestock has its own requirements of soil and climate, labour and other resources, and analysis of individual components is further complicated by the interdependence of different enterprises. Crops may be grown or livestock kept as much for the indirect contributions they make to the whole economy of the farm as for the revenue they provide; and, while this is becoming less marked as farming systems are simplified and farmers become more cost-conscious, there is still much validity in the observation by Astor and Rowntree that 'it is almost true to say that in British agriculture nothing is ever done for its own sake alone'.

Furthermore, the motives of farmers are very varied. While farming is primarily an economic activity, as the sympathetic movements of production and prices witness, farmers are also influenced by their views of what constitutes an acceptable standard of living, by the value they place upon their independence, by their respect for traditional practices and for the opinions of their neighbours, by their knowledge of the alternative enterprises open to them and by their technical competence to undertake them. Few studies have yet been made of the importance of these considerations in influencing farmers' decisions, but J. B. Butler's *Profit and Purpose in Farming* provides a valuable insight into the variety of motives and attitudes and their effects upon the choice of farming systems.

Even when economic motives are dominant, few farmers knowingly achieve profit maximisation, and many have little idea how nearly their present farming systems approach it.

INTRODUCTION

Farmers are certainly better informed on economic matters than formerly, the keeping of farm accounts is now encouraged, and the agricultural press, the radio, the farmers' unions, the marketing boards and the advisory services provide abundant information on prices, costs and profits. Nevertheless, many farmers lack essential information to evaluate costs and profits or hold erroneous views about them; for example, sample studies of wheat and barley have shown that many farmers could not give an accurate estimate of their yields even after they had disposed of their crops, and an enquiry into herd replacements in northern England found that farmers greatly over-valued the economic advantages of breeding dairy herd replacements. Difficulties also arise where costs are shared among several enterprises, where produce is consumed mainly on the farm and there is no realistic market price, or where there are indirect benefits to other enterprises or to the whole farm economy.

Of course, farmers have no absolute freedom in choosing enterprises, even when they are adequately informed about what is most profitable. In the short run at least, farmers cannot choose a farm appropriate to the enterprises they wish to pursue: normally they must select from among those enterprises for which the size, layout, physical features and fixed equipment of their farm are suited, and these characteristics can be changed only slowly by the acquisition of more land, by land improvement, by rationalisation of layout and by the erection or modification of farm buildings. Inputs of labour, machinery and fertilisers can more easily be varied, but even these depend partly on such considerations as farm size, soil, climate and accessibility. For example, it is unlikely that the present distribution of agricultural labour can be greatly altered in the forseeable future, and enterprises requiring large inputs of labour will be impracticable in remote areas of prolonged rural depopulation, where social services are scarce and standards of housing poor. Field size and slope may exercise similar constraints on mechanisation.

Where the attempt is made, as in this book, to explain, not the actions of individuals, but what large numbers of farmers tend to do, further problems arise. Even if measurement of the profitability of each enterprise were simple and an enterprise could be shown as highly localised in a region, no suitable data from which any regional advantage could be demonstrated exist for a sufficiently large sample of farms. The farmers who keep accounts for the Farm Management Survey are a small proportion of the total population of farmers and are neither a random nor a systematic sample, so that conclusions about small areas could not safely be drawn from their records, even if such data were available. Other samples are generally selected with probability proportional to the size of enterprise and, while this is satisfactory where the aim is to study the enterprise as a whole, it cannot be used to establish regional differences in comparative advantage; in the National Wheat Survey, for example, where most of the sample farms were located in eastern England and only a few in western counties, the data will not permit valid conclusions to be drawn about the relative suitability of these areas for wheat growing.

In these circumstances, one may attempt to establish comparative advantage subjectively, either by reference to the intrinsic suitability for each enterprise of different environments, including both natural and man-made features, or by examining what farmers in fact do. The latter approach needs care, for it may involve a circular argument: we infer that a particular enterprise enjoys a comparative advantage in an area because this is the enterprise that most farmers have chosen, and then seek to explain its location in terms of its comparative advantage, although the reason may be related to some traditional practice, to chance or to some other factor. The first approach is the more logical, but there is no objective method of comparing relative suitability for different enterprises, although estimates of physical output may provide some insight where they are available in comparable terms, as with yields of cereals.

THE SCOPE OF THE BOOK

These considerations have influenced the approach adopted in this book and the arrangement of the chapters. The following chapter puts into perspective the changing context, national and international, in which British farmers have made their decisions over the past 150 years, during which the

United Kingdom ceased to be largely self-sufficient in food and became a major food importer; for although there has been an improvement since the 1930s, when home production supplied only a third of the country's food requirements, competition from the produce of other, often better endowed, lands remains a major concern of British farmers. In this period, too, the United Kingdom changed from a predominantly rural to an overwhelmingly urban country, and agricultural policy has swung from protection to laisser-faire, and finally to government involvement in virtually every aspect of British farming. Adequately to cover this field would require several volumes, and only the broadest outlines can be sketched in a single chapter; thereafter these external constraints on British agriculture must largely be taken for granted.

The following four chapters attempt a systematic examination of major factors affecting choice of enterprise, viz. land (in the widest sense), labour, machinery and markets. Ultimately each of these can be modified to varying degrees, but they circumscribe a farmer's immediate freedom of choice. The treatment is limited by the data available, but an attempt is made to show how quality of land, size of farm and availability of men, machines and markets vary throughout the country, and to indicate briefly, in anticipation of the study of individual enterprises, the relevance of these considerations to the varied pattern of farming in Great Britain.

The bulk of the book is devoted to the principal enterprises undertaken on British farms. Particular attention is paid to the spatial characteristics of each enterprise and, where possible, to its relationship to other similar enterprises; but the treatment depends on the nature of available data. Few of these enterprises have been studied by geographers, and none more recently than 1960; furthermore, despite the numerous enterprise studies published by agricultural economists, there are few country-wide surveys of individual enterprises. A notable exception is the Britton report on *Cereals in the United Kingdom*, which is a mine of information about the regional pattern of cereal production, though this was not the main purpose of the survey. Agricultural economists at Wye and Cambridge have made available much of the material necessary for a geographical study of horticulture, the milk marketing boards have conducted numerous surveys of dairying, and developments in poultry farming have also been examined; but for beef cattle, sheep and pigs one must rely heavily on the agricultural census and on a very diverse literature, little of which considers the location of these enterprises. The resulting treatment is inevitably somewhat uneven.

While the complexity of British agriculture necessitates approaching its geography through the study of individual enterprises, the ultimate objective is an understanding of how these different enterprises fit together to form recognisable farming types which vary in both absolute and relative importance throughout the country. Important as this topic is, lack of data again makes possible only a preliminary view. Only one chapter is devoted to it and, since the published material on enterprises and types of farming relates mainly to large, heterogeneous regions, the treatment of types of farming is somewhat generalised. Attention is focused on the problems of identifying types of farm and farming regions and on the importance of scale in determining acceptable generalisation; for a compromise must be struck between the extremes of over-simplification that hides significant differences and regional divisions so closely approaching the complexity of the real world that they do little to clarify any underlying relationships.

What is not attempted in this book is a detailed analysis of British agriculture, region by region, similar to Wilfred Smith's survey in his *Economic Geography of Great Britain*, for this would have lengthened the book considerably and have involved further repetition of material already examined systematically. Furthermore, while there is certainly more material available than when Smith wrote, the rate of agricultural change is now much faster; even many regional accounts of agriculture which have appeared in the post-war handbooks of the British Association and in similar publications are now out-dated. It seems increasingly clear that the only satisfactory approach to the detailed study of agricultural regions in Great Britain must be a historical one that attempts to answer the question 'How does an agricultural region originate?', and is therefore beyond the scope of the book.

Agricultural geography is properly regarded as a branch of

economic geography, for agriculture provides a living for most of mankind and, in developed countries at least, farmers' motives are primarily economic; yet, for reasons already discussed, this book is almost innocent of economic data, as are most texts dealing with the geography of agriculture.

Some agricultural economists, while recognising that the major physical contrasts between the north and west and the south and east do influence the type of farming, regard the search for any greater degree of spatial order in a country as small as Great Britain as misguided; for they have shown that the range of profitability among farms with the same type of farming is often greater than the mean range of those with different types of farming. Geographers, while acknowledging the great importance of management in accounting for inter-farm difference, retain a conviction that the distribution of farm-types and enterprises is less haphazard than economists suggest, that the regional differences in the factors of production that have been shown to exist must affect both the efficiency and the profitability of farming, and that more sophisticated data would reveal smaller intra-class and wider inter-class differences than are now apparent from the use of large regions. Unfortunately, the available samples are too small either to substantiate or to disprove these convictions, and analyses of large heterogeneous regions may be quite misleading. Yet, even if differences in natural features, farm layout, labour supply, fixed equipment and accessibility have less effect than differences in management ability, the increasing regionalisation of agriculture in recent years suggests that, with the greater competitiveness of the past decade, farmers have become more conscious of their importance in their choice of farming systems.

It has been suggested that every historical situation poses two questions, 'What did they think they were doing?' and 'What did they really do?' Similar questions can be asked in a geographical context, but in this study 'they' are tens of thousands of farmers who rarely give explicit reasons for their choice of farming systems and, indeed, may have no clear ideas about the basis of their decisions. Attention is therefore best directed to discovering what farmers in fact do and have done; when this has been satisfactorily answered and the results analysed, we may gain more insight into the reasons why.

GENERAL

More complete bibliographies will be found in W. Smith, *An Economic Geography of Great Britain* (1948), which covers the period to 1939, and J. T. Coppock, 'Post-war studies in the geography of British agriculture', *Geogr. Rev.* 54 (1964), pp. 409–26.

Other sources covering many aspects of the agricultural geography of Great Britain include:

J. T. Coppock, *An Agricultural Atlas of England and Wales* (1964)

J. G. S. and P. Donaldson, *Farming in Britain Today* (1969)

F. H. Garner (ed.), *Farming Systems in Britain* (in press)

L. D. Stamp, *The Land of Britain: its Use and Misuse* (3rd edition, 1962)

G. P. Wibberley, *Agriculture and Urban Growth* (1959)

H. T. Williams, *Principles for British Agricultural Policy* (1960)

Central Office of Information, *Agriculture in Britain* (1969)

Statistical sources include:
K. E. Hunt and K. R. Clarke, *The State of British Agriculture 1965–6* (1966)

Department of Agriculture and Fisheries for Scotland, *Agricultural Statistics Scotland*, annually

Department of Agriculture and Fisheries for Scotland, *Scottish Agricultural Economics*, annually

Milk Marketing Boards, *Dairy Facts and Figures*, annually

Ministry of Agriculture, Fisheries and Food, *Agricultural and Food Statistics*, Guide to Official Sources, No. 14 (1969)

Ministry of Agriculture, Fisheries and Food, *Agricultural Statistics, the United Kingdom*, Parts I and II

Ministry of Agriculture, Fisheries and Food, *Agricultural Statistics England and Wales*, annually

Ministry of Agriculture, Fisheries and Food, *Farm Incomes, England and Wales*, annually

Ministry of Agriculture *et al.*, *Annual Review and Determination of Prices*, annually

General discussions relevant to the field of agricultural geography will be found in:

R. O. Buchanan, 'Some reflections on agricultural geography', *Geography* 44 (1959)

J. B. Butler, *Profit and Purpose in Farming* (1960)

J. T. Coppock, 'The geography of agriculture', *J. Agric. Econ.* 19 (1968–9)

R. J. C. Munton, 'The economic geography of agriculture', in *Trends in Geography*, edited by R. U. Cooke and J. H. Johnson (1969)

CHAPTER 2

The Changing Context of British Farming

Although in broad outline the agricultural geography of Great Britain, epitomised by the contrasts between a mainly pastoral west and a predominantly arable east, shows considerable stability, the network of internal and external relationships that provides the context within which farmers make their decisions has changed fundamentally since the end of the eighteenth century. From a rural society, largely self-sufficient in food, Great Britain has become the most highly urbanised and one of the most densely peopled countries in the world, with a population of 55 millions uniquely dependent on imported agricultural produce. The effects of population growth and urbanisation, the main instruments of change, have been complicated by a revolution in transport and by rising standards of living and, since the 1930s, by increasing government intervention in agricultural affairs, as well as by changes within agriculture itself. This chapter presents a brief account of this changing context as a prelude to examining the present agricultural geography and recent changes in it.

URBANISATION & POPULATION GROWTH

At the end of the eighteenth century, Great Britain had a population of only 10·7 million inhabitants, or less than a fifth of the present population, and was still predominantly rural, although industrialisation and urban growth were rapidly changing the balance between town and country, and agriculture already employed less than half the working population. Before the repeal of the Corn Laws in 1846, wheat and sugar were the only major foodstuffs imported on any scale, accounting for 62 per cent. of agricultural imports by weight and 79 per cent. by calorific value in the 1830s, while British livestock depended almost entirely upon grazing and home-grown fodder crops. Yet, though wheat was a major item among food imports, 90 per cent. of the country's requirements were still home grown, and it has been estimated that British agriculture provided nearly 95 per cent. of the calories required by the population of the United Kingdom (then including the whole of Ireland). Internal transport was largely by cart, although the distances over which produce moved to urban markets were often surprisingly large, and cattle travelled hundreds of miles on the hoof from Wales and Scotland to be fattened nearer London.

By 1851 the population, then totalling 20·9 millions, seems to have been almost equally divided between urban and rural. Thereafter, the urban population grew rapidly, to nearly 80 per cent. of the total population of 40·9 millions by 1911; moreover, the urban population was being increasingly concentrated in large cities, especially the seven conurbations. Over most of rural Britain migration to the towns and overseas was leading to rural depopulation in the second half of the nineteenth century, although later this trend of falling population was generally reversed around the rapidly expanding urban settlements, as more of those who worked in towns and cities sought to live in a rural environment. The relative rise in rural population since 1931, however, is in part illusory, resulting from the failure of administrative boundaries to change with urban development; for the true proportion of urban dwellers is now probably nearer 90 per cent. than the 80 per cent. indicated by the population census.

Agriculture in Great Britain is consequently dominated by the needs of a largely urban community and agriculture employs only a small proportion of the working population. Even by 1801 agricultural workers were in a minority and agriculture's share has fallen from an estimated 20 per cent. of the employed population in 1851 (25 per cent. of adult workers) to under 4 per cent., the lowest proportion in any major industrial country, partly because industrialisation began earlier here and because Great Britain, unlike most of continental Europe, has not retained a strong peasant agriculture. Even in the strictly rural areas agriculture's share of employment has declined as living standards have risen and an increasing proportion of the population is employed in the service sector.

The growth of population on a land area which is essentially fixed has meant a fall in the area available for food production per head, accentuated by the accompanying loss of agricultural land, generally of above-average quality, to urban and other uses. Between 1850 and 1900 the urban area probably increased by over a million acres and since 1900 by a further $2\frac{1}{2}$ million. The rate of loss has been affected not only by population growth but also by improvements in living standards and by the need to rehouse those living in substandard accommodation at high densities. The adoption of lower housing densities after the First World War, combined with the depressed state of agriculture and the consequent willingness of owners to sell, led to large losses of land between the two World Wars, at an average rate of over 60,000 acres a year in the 1930s. An increasing range of other demands on land for communications, mineral working, water storage, recreation and the like also contributed, although strict planning control and a policy of conserving good agricultural land wherever possible has moderated the scale of losses since the Second World War. Besides these losses more than a million acres, mainly of rough grazing, have been taken for forestry since the First World War, and such losses averaged as much as 40,000 acres a year in the mid-1950s. Between 1850 and 1939 the agricultural land available per head of population probably fell from nearly $2\frac{1}{2}$ acres to under one acre, or from about 1·6 to 0·6 acres of crops and grass. Even between 1939 and 1965 it fell by more than 10 per cent. as the population increased and the area of agricultural land diminished.

The effects of urbanisation and growth of population were reinforced by other changes. Living standards are notoriously difficult to measure, but they certainly increased considerably over this period. Between 1850 and 1900 average real wages rose by about four-fifths and between 1900 and the outbreak of the Second World War by a further 30 per cent. Since 1945 they have risen steeply by more than a half, although considerable regional disparities in average income remain, notably between the London area and the far north.

With these changes in spending power have come changes in

patterns of food consumption, although here too there are considerable regional differences. Consumption of bread and potatoes per head has diminished and that of meat, dairy produce, fruit and vegetables has risen. From the early 1860s to the eve of the First World War per capita consumption of wheat fell by 16 per cent., while that of meat rose by 35 per cent.; similarly, between 1880 and 1911–13 consumption of potatoes fell by 24 per cent., while that of butter rose by 33 per cent. There were further changes between the two world wars, with per capita consumption of meat rising 48 per cent. above the 1861–5 level and that of butter 108 per cent. above the 1880 level, while consumption of fruit and vegetables also increased by 87 per cent. and 110 per cent. respectively between 1911–13 and 1936. Since the 1930s per capita consumption of milk has risen by 57 per cent. and that of poultry meat by 223 per cent. The amount of fresh fruit and vegetables eaten has declined, while that of frozen, canned and bottled produce has increased, an interesting effect of changing technology. Such changes in diet have obvious implications for agricultural production.

TRADE & TRANSPORT

The growth of population and rising standards of living, especially the trend towards increasing consumption of livestock products, which require more land to support the same number of people than does crop production, would themselves have implied an increase in food imports in the absence of any sharp rise in yields. In fact yields, already high by European standards, showed only a small upward trend until the Second World War, but the volume of imported foods increased more rapidly than the shortfall in domestic production and competition from imports itself became a major instrument of agricultural change in its own right (Figure 1).

The first impact of increasing imports of foodstuffs was felt by those farmers who depended on wheat as a cash crop. Certainly, the repeal of the Corn Laws did not immediately have the effects some critics had forecast, largely because trade was disrupted by the American Civil War and the Crimean War, but the expansion of the railway network in North

America and improvements in trans-Atlantic shipping soon led to rapidly increasing imports of wheat, the effects of which were accentuated by a succession of bad harvests and by financial crises. British farmers faced competition from overseas territories where natural conditions were often better suited to grain production, and, unless cereals were to be kept out by

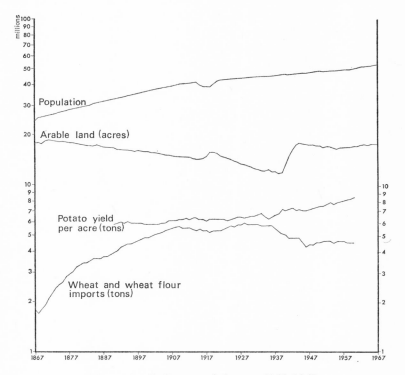

Figure 1a. Indicators of change, 1867–1967
Sources: Annual Abstract of Statistics and agricultural censuses. Note that imports are for the United Kingdom; all other data relate to Great Britain. Only potato yield is measured on the right hand scale.

tariffs, their farming system had to be changed. Imports of wheat doubled between 1850 and 1872; thereafter they rose from some 2 million tons to over 5 million tons in 1910 (Figure 1a). Moreover, since demand was rising and home production declining with falling prices, the share of total supplies contributed by imports rose from 25 per cent. in 1852–9 to 55 per cent. in 1876–8.

Although cereals were the first product to experience strong overseas competition, there were soon others, particularly after the development of refrigeration in 1882 allowed meat and dairy produce to travel half way round the world from Australia and New Zealand. Meat imports rose fourteen-fold between 1861–5 and 1911–13, while butter and cheese imports increased four-fold and three-fold respectively between 1861–5 and 1911–13. Only those products, such as fresh milk and potatoes, where perishability or bulk provided natural protection, were exempt. Large quantities of animal feeding stuffs were also imported: imports of maize quadrupled between 1867–9 and 1894–1903, to reach nearly $2\frac{1}{2}$ million tons. By 1909–13, therefore, the United Kingdom was importing 79 per cent. of its requirements of grain and flour for human consumption, 40 per cent. of its meat, 72 per cent. of its dairy produce, 73 per cent. of its fruit and 62 per cent. of its animal feeding stuffs. British farmers now had to face the competition of farmers who, through specialisation and more favourable natural conditions, could overcome the barriers of distance and still undercut them in British markets. This challenge was also faced by other European countries, but these, with large peasant populations, generally erected tariff barriers to keep imports out. In the United Kingdom, the arguments for cheap food were accepted, to the benefit of the nation as a whole, and British farmers received no protection and little other help, although the adjustments forced on them were perhaps ultimately advantageous for the structure of British farming.

During the First World War, imports were greatly restricted and the volume of home produce increased by 25 per cent.; but reliance on overseas supplies was resumed after the war and only 30 per cent. of the United Kingdom's annual requirements of food were home-produced in the late 1930s. This dependence was again reduced during the Second World War, when the volume of imports was halved; at the same time there was a great increase in home production, and net output (measured in calories) was 91 per cent. higher in the peak year of 1943 than pre-war. Since 1945 improved agricultural productivity has further reduced the dependence on overseas supplies, despite a rise in population of over six millions, and the United Kingdom now produces about half the food it requires and two-thirds of what can be grown in temperate climates. The dependence on wheat and wheat flour has lessened from 77 per cent. to 53 per cent. and on carcase meat and offal from 49 per cent. to 30 per cent., the country is now self-sufficient in eggs and, although sugar and cheese are still mainly imported, the share of home production has almost doubled. Butter, oils and fats remain the products where dependence on overseas sources is greatest.

It was above all improvements in transport which, by lowering costs and extending the distance over which perishable produce could be carried, enabled distant producers to compete on favourable terms with British farmers, 'with transport costs little heavier between continents than they had once been between counties' (J. H. Clapham). Costs fell dramatically with improvements in shipping and railways; wheat, for example, could be transported half way across the United States and shipped across the Atlantic to British ports for only 2s. 10½d. a bushel in 1902, compared with 11s. in 1868–79. The growth of railways in overseas territories extended the area from which crops and livestock products could be sent to Europe and, more recently, air transport has permitted perishable luxury products to compete in British markets. Technical improvements in transport and in the treatment of produce have also enabled more commodities to be transported over long distances. Chilling of beef, for example, reduced the disadvantages that imports from Argentina suffered by comparison with the quality of home-killed produce, although it was not until after the Second World War that these benefits extended to Australia and New Zealand. Quick-freezing and dehydration have made increased imports of vegetables possible and, more recently, the development of long-storage milk has potentially extended the distance over which milk for liquid consumption can be transported, although the effect of this advance has not yet been felt for other reasons.

Developments of transport within Great Britain, especially the extension of the railway network, which increased from nearly 5,000 miles in 1848 to over 14,500 in 1875, also had considerable consequences, enabling perishable produce such as milk to be brought from further afield and permitting areas such as west Cornwall to exploit natural advantages for the production of early crops which their remoteness had previously denied them. In eastern arable districts light railways were specially constructed for the transport of farm produce. The railways could also transport livestock and meat more swiftly so that, where other conditions permitted, areas which formerly depended on store cattle, such as Aberdeenshire, could fatten them.

The internal combustion engine has further minimised the effects of proximity to markets, particularly since the Second World War. At first, the role of road transport was to act as feeders to the railways, but the lorry has increasingly supplanted the goods train as the prime mover of agricultural produce, little of which now goes by rail. As we shall see later, rail traffic in livestock has fallen sharply since 1950 and is likely to contract still further as the number of loading points is drastically reduced. The flexibility of road transport, especially in size of load and choice of route, has real advantages in agriculture, where produce must be collected from many small producers over a large area. Locational advantages tend to have marginal rather than primary effects, although much depends on the bulk and the value of the produce. The importance of location is further reduced by transport pricing policies which fix uniform charges over large areas, so that individual producers do not pay the true costs of transport, which are either equalised among all producers or passed on to the consumer. Indeed, immediate proximity to towns may now have disadvantages because of the greater risk of trespass and damage and the heavier demands for land for non-agricultural purposes.

It was, however, the developments in international transport which were decisive for British farmers as a whole and which encouraged them to concentrate on those products for which they had locational and natural advantages: on grassland farming, to which the British climate is well suited; on the rearing and fattening of livestock, where labour costs were comparatively low and the home-killed product had advantages in quality over imported chilled and frozen meat; on liquid milk, where the home producer had, until recently, a natural monopoly, provided by distance, with only imported condensed

THE CHANGING CONTEXT OF BRITISH FARMING 19

or evaporated milk as a rival; on fruit and vegetables, where freshness and quality were also important; and on bulky goods, such as potatoes, where transport costs were high in relation to value. True, over much of the country the choice of alternative enterprises was limited. Whatever the price of wool

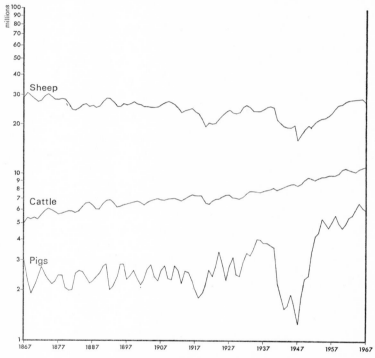

Figure 1b. (See figure 1a for sources).

or of imported New Zealand lamb, there was no alternative to sheep rearing in many upland areas, and farmers either kept sheep or abandoned farming. Similarly, the possibilities in northern and western parts of Great Britain were much more restricted than in the south and east, although, with a long-established emphasis on livestock farming, they had less need to change.

There was consequently increasing emphasis on livestock farming, and E. Ojala has estimated that, whereas the gross output from livestock was 20 per cent. greater than that from crops in 1860–9, it was 168 per cent. larger in 1911–13, the output from crops having halved in the interval and that from livestock having risen slightly. Numbers of cattle increased and the acreage under tillage crops declined steadily (Figure 1).

These changes in British farming continued after the First World War, with the emphasis increasingly on milk and other perishables: the volume of dairy produce increased by about 14 per cent., of poultry and eggs by 38 per cent. and of fruit by 25 per cent. between 1908 and 1925. Nevertheless, the advantages of proximity enjoyed by British farmers were being weakened and, but for major changes in policy, particularly the abandonment of free trade and the increasing intervention by government in agricultural affairs, British producers of milk, fruit and vegetables would have undoubtedly felt increasing competition from European farmers who enjoyed natural advantages over them. Already, Dutch dairy farmers have been able to offer lower tenders for the supply of milk to American forces in Great Britain, and the lower production costs of Italian horticulturalists more than offset higher transport charges. Without tariff barriers, quotas and voluntary agreements with exporters to restrict the volume of produce sent to Great Britain, and without the benefits of collective marketing, price supports and other forms of governmental assistance, British farmers would have had to make other substantial adjustments.

GOVERNMENT INTERVENTION

Unlike most other European countries, Great Britain had imposed few restraints on imports of agricultural produce, beyond those required in the interests of plant and animal health, and had offered little other help to farmers than the partial derating of agricultural land. In the early 1920s, to promote home production of sugar and to help arable farmers, the government remitted excise duty on home-grown sugar and the British Sugar (Subsidy) Act of 1926 provided a subsidy, intended to be temporary, to producers of sugar beet. With the world depression of the early 1930s, however, commodity prices fell sharply and, for the first time since the repeal of the Corn Laws, British governments provided some protection for

British farmers and began to modify the policy of laisser-faire by providing subsidies or guaranteed prices to producers, by facilitating the regulation of internal marketing and, from 1937, by grants to farmers to encourage desirable practices.

Although the United Kingdom remained the major food importer in the world, the abandonment of free trade led to a reduced volume of food imports. These had risen by 17 per cent. between 1927–9 and 1931, when world food prices slumped, but they declined by 13 per cent. between 1931 and 1934 and remained at about this level until 1939. Tariffs on agricultural imports were mostly at rates between 10 and 20 per cent. *ad valorem*, although those on fruit and vegetables varied seasonally, being higher during periods when the home crops were being marketed. With the exception of hops and sugar, however, these duties applied only to foreign produce; under the Ottawa Agreements of 1932, Empire produce continued to be admitted duty free and there was a shift from foreign sources of food imports to Empire sources, whose share rose by 42 per cent. Voluntary agreements were also negotiated with exporters of beef, lamb, mutton, eggs and bacon to restrain their exports, some of these being later replaced by compulsory limitations on imports; imports of potatoes were similarly regulated. Such quantitative restrictions were not re-imposed after the Second World War, but during the 1960s attempts have again been made to negotiate similar agreements, to restrain imports and even to seek higher prices in order to raise the level of domestic prices.

Horticultural producers undoubtedly benefited from the restrictions on imports, but it is not clear how far these measures improved the economic position of British farmers. More effective for agricultural production in Great Britain have been the other measures taken by governments. Until 1931, the role of government had largely been restricted to regulating crop and livestock hygiene, to encouraging agricultural research in a modest way and to modifying tenurial relationships, by passing legislation to enlarge the rights of tenants to compensation for improvements made during their tenancy and to encourage the establishment of statutory small holdings. A major change was the Wheat Act of 1932, under which producers of wheat received a subsidy of the difference between average prices received for British wheat and a standard price, although the deficiency payment was reduced proportionately as sales of wheat increased beyond a specified quantity. Under an Act of 1934 a subsidy on fat cattle was paid irrespective of selling price. During the 1930s, guaranteed minimum prices were introduced for barley, oats, bacon pigs, sheep and milk for manufacture into butter and cheese. Most of these schemes ended too soon for their effects to be judged, but the Wheat Act undoubtedly stimulated domestic wheat production; between 1932 and 1933, when acreage under the plough was declining, the wheat area rose by 400,000 acres, or 30 per cent.

Other measures were taken under the Agricultural Marketing Acts of 1931 and 1933. These facilitated the establishment of producer-controlled marketing boards to strengthen the buying power of producers and to promote orderly marketing; boards were established for hops, milk, pigs and potatoes. Their functions, however, varied considerably. The Hop Marketing Board, the first to be established, purchased all hops sold and restricted production by acreage quotas, given to those farms where hops were being grown when it was established; its effectiveness was greatly strengthened by heavy duties and strict quotas on imported hops. The four milk marketing boards (one for England and Wales and three for different parts of Scotland) equalised the prices received by producers for liquid milk and for manufacturing milk, which had suffered from the great rise in imports of dairy produce, so removing the danger that producers of milk primarily for manufacturing would undercut the liquid milk market; these boards also attempted to enlarge producers markets by encouraging the consumption of liquid milk. The Pig Marketing Board was concerned with bacon pigs only and attempted to negotiate contract prices for its members; chiefly because of the elastic demand for bacon and of the large imported supplies of bacon, it was less successful than the other boards, which dealt primarily with commodities enjoying natural or man-made protection, and it was not revived after the Second World War. The Potato Marketing Board also attempted to control production by means of acreage quotas, with fines for producers who exceeded their allotment, and by regulating the size of potatoes sold; it, too, was suspended during the Second World

War and not revived until 1955. These boards undoubtedly helped to maintain the prices received by producers, although only the milk marketing boards have been a major cause of changes in the geography of agriculture.

A third major change of policy was the introduction of modest grants to encourage desirable practices and to increase the productivity of farming in general. Since the 1870s, the agricultural industry had been depressed. Because rents had fallen greatly, landlords lacked resources to maintain fixed equipment, while low prices made it impossible for farmers to practise high farming. Consequently, land was neglected, under-draining deteriorated and the soil was starved of lime and plant nutrient. The 1937 Agriculture Act provided subsidies on lime and basic slag to reduce their cost to farmers, and in 1939, with war imminent, a subsidy was authorised for ploughing up permanent pasture.

In general, these measures were introduced piecemeal to meet a crisis rather than as parts of a coherent and consciously-formulated agricultural policy, but the outbreak of the Second World War made it necessary to expand domestic food production to compensate for the inevitable reductions in imports of food and feeding stuffs. Using experience gained in the First World War, the Government intervened more directly in British farming, setting production targets and encouraging greater output by price incentives, by acreage payments on wheat and potatoes and by subsidies on hill sheep and hill cattle and, where necessary, by prescribing cropping. They also aided rehabilitation of farmland by continuing grants for ploughing permanent grassland, by introducing grants for drainage improvements and land reclamation, and by subsidising fertilisers. They set up County War Agricultural Executive Committees, who issued cropping directives, provided a channel for the implementation of policy and even took over and farmed land which had been neglected. Most of the functions of marketing boards were suspended and, through the Ministry of Food, the Government became the main purchaser of agricultural produce. As a result, the index of gross agricultural production (measured in calories) rose by 55 per cent. between 1938 and 1943 and, since imports of feedingstuffs were curtailed by as much as 85 per cent., the proportionate increase in net output was even greater at 91 per cent. In part this achievement represented an enforced change in diet, and a comparison by value is much less favourable to war-time achievements, the indices of output rising by only 4 per cent. and 15 per cent. respectively.

With the food shortage which prevailed at the end of the Second World War and with balance of payments difficulties, it was still necessary to encourage agricultural production and targets continued to be set, with the object of achieving an output some 55 per cent. above pre-war level. Food rationing was retained and the Ministry of Food remained a major buyer of foodstuffs. More important for the future of farming in Great Britain was the acceptance of permanent state support for British agriculture and the formulation, for the first time without threat of war, of a coherent policy to promote, in the words of the preamble to the 1947 Agriculture Act, 'a stable and efficient industry, capable of providing such part of the nation's food as in the national interest it is desirable to produce in the United Kingdom, and of producing it at minimum prices consistent with proper remuneration and living conditions for farmers and workers in agriculture and an adequate return on capital invested in the industry'. This policy was effected chiefly through guaranteed prices for the major agricultural products, viz. fat cattle, fat pigs, fat sheep, eggs, milk, wool, cereals, potatoes and sugar beet, and through production grants similar to those introduced before and during the war; because of the variability in home supplies, support for horticulture was continued by regulating supplies of competing imports. Each year the ministers responsible for agriculture in the constituent countries of the United Kingdom are required to review the economic conditions of the industry with representatives of producers. In the light of this review, ministers determine the level of guaranteed prices for the next harvest or for the following twelve months for livestock and livestock products.

From the mid-1950s, as the world food situation improved and problems of agricultural surpluses arose, the emphasis in agricultural policy switched from maximum production to economic production. Rationing was ended in 1954, food marketing was returned to private traders and to the marketing boards, and new boards were established for eggs, wool, tomatoes and cucumbers. To discourage excess production, standard quantities, to which alone price guarantees applied, were introduced for eggs, milk, barley, potatoes, sugar beet and wheat; where production exceeded the standard quantity, the effect was to lower the average price received by all producers. On the other hand, the 1957 Agriculture Act limited changes in the level of guarantees (although not, because of the introduction of standard quantities, in that of prices received by farmers). Under this Act, the total value of guaranteed prices and 'relevant production grants', i.e. other than grants for long-term improvements, may not be less than $97\frac{1}{2}$ per cent. of that in the preceeding year, after allowing for changes in costs; furthermore, the reduction in the guaranteed price may not be greater than 4 per cent. per annum for any individual crop, or 9 per cent. over three years for any livestock or livestock product. The Act also abolished the powers, inherited from the Second World War, to supervise and dispossess farmers.

Since the 1947 Agriculture Act, agricultural produce has generally been allowed unrestricted entry and prices have been allowed to find their own level, payments under the price guarantees being related to the differences between market price and guaranteed price, in contrast to the policies of the European Economic Community, where high prices are maintained through import duties. Since the early 1960s, however, fluctuations in the volume of imports and in domestic prices (and hence in exchequer payments) have encouraged governments to seek agreements with overseas producers to secure voluntary quotas, a reversion to the policy adopted in the 1930s. Such arrangements were negotiated for bacon in 1964 and cereals have also been the subject of agreement with exporting countries. The Home Grown Cereals Authority and the Meat and Livestock Commission were also established to encourage orderly marketing of the principal home produce for which there were no marketing boards. Grants have been much more important in Government aid for agriculture during the post- than in the pre-war period, although their share of the total value of exchequer support has varied considerably. Some of the grants available support particular enterprises, notably in problem areas, as with the hill sheep and hill cattle subsidies, but some are more general, as with the calf and beef cattle subsidies; some encourage desirable practices, as with the subsidies on lime, fertilisers and silo construction; and some encourage long-term improvements in agriculture, notably those authorised under the Farm Improvement Scheme for the improvement of fixed equipment and also for the amalgamation of holdings. A similar Horticultural Improvement Scheme was introduced in 1960. In recent years, the share of exchequer production grants has tended to rise and that of implementation of price guarantees to fall; thus, between 1955–7 and 1966–8 the proportions altered from 30 per cent. and 70 per cent. respectively to 48 per cent. and 52 per cent. (compared with 8 per cent. and 92 per cent. in 1938–9). The nature of grants payable has also varied; for example, ploughing grants were abolished in 1949, reintroduced in 1952, reduced in 1963 and abolished again in 1967, and the basis for grants for improvements to hill land has changed several times.

Other government support for agriculture and horticulture has also increased. Attempts have been made to improve the quality of produce, through measures imposed in the interest of hygiene, as with dairy regulations, or to encourage grading of produce, as with the statutory schemes for apples and pears. Advisory services, provided free to the farmer, have been greatly extended since the 1930s and now include advice on capital investment and farm accounts as well as on agronomic matters. State-supported agricultural research has also increased greatly.

Government influence on agriculture goes beyond overtly agricultural policies. Fiscal measures often have important side effects for British farming; thus the scope for offsetting business profits against farming losses, which existed before the 1961 Finance Act, led to a flow of funds into agriculture. Concessions on estate duty favouring investment in agricultural land and planning policies giving priority to safeguarding good agricultural land where possible or placing restraints on agriculture in the interests of amenity are other examples of policies with important consequences for agriculture.

It is difficult now to envisage politically acceptable policies which would end government involvement in agriculture. Debates are more likely to centre on methods of government intervention or on the scale of support, and entry into the

European Economic Community would entail substantial changes in both the methods and the levels of intervention. The consequences of such intervention are, however, more debatable, for some have been unintentional or incidental, and their impact on individual commodities will be considered in subsequent chapters. Since the policy accepted in 1947, agriculture has been more prosperous and its returns more certain than in the laisser-faire period before the 1930s; but equally the 1958 Agriculture Act, which defined the properly payable rent as the open market rent, assisted the steep rise in rents and land values that has greatly increased pressures on farmers' profits. Given the continued dependence of the United Kingdom on food imports and the emergence of agricultural surpluses in developed countries, producers of cash crops and livestock products would undoubtedly be even harder pressed without price supports, statutory marketing schemes and controls over imports; for direct and indirect aid to agriculture represents over 50 per cent. of net farm income since the passing of the 1947 Agriculture Act and was 80 per cent. in 1961–2. Even this proportion is an understatement because the prices maintained by marketing boards, notably in milk and potatoes, are higher than would obtain in free market conditions, so the true figure was probably about 90 per cent. in 1960. On some types of farm grants and subsidies may occasionally even exceed net farm income; for example, in 1965–6 hill sheep, hill cattle, calf and winter keep subsidies represented 109 per cent. of net farm income on a sample of upland farms in Scotland. Another way in which the external relationships of agriculture have changed over the past hundred years is the increasing interdependence of agriculture and manufacturing industry. 'Before 1912 farming found its own balance' (H. G. Sanders), but the great expansion in inputs of fertilisers and pesticides and the great increase in mechanisation, discussed in Chapters 3 and 4, with the corresponding decline in the importance of farmyard manure and horse traction, have greatly altered the proportion of inputs derived from the farm itself. Some £800 million is paid by farmers to non-farmers for goods and services, a sum equivalent to nearly half the gross output of British farms. Similarly, over 60 per cent. of the agricultural produce used for food now undergoes some kind of processing. Farm produce is thus increasingly raw material for the food industry and for other branches of manufacturing industry and farming is a major market for the products of industry. Thus, although agriculture's share of the Gross National Product has declined from 17 per cent. in 1867–9 to 6 per cent. in 1911–13 and 3 per cent. in 1966–8, agriculture's links with the economy have become more complex and agriculture is increasingly only one stage in a much more extended production process.

Competition from farmers in other countries has long affected the character of British farming; now growing demands for land for other purposes are increasingly affecting it also. Upland agriculture has had to share land with grouse and deer and now with the armed services, with water authorities and, increasingly, with recreationists, all imposing some restraint on the way land is used for farming. At the same time, government has become progressively involved, both directly in determining minimum prices and indirectly by promoting, whether intentionally or not, numerous changes in agriculture. Throughout this book these external relationships will be taken for granted; they should not be forgotten.

FURTHER READING

Much writing in agricultural and economic history bears on the theme of this chapter, especially J. H. Clapham *An Economic History of Modern Britain*, 3 Vols. (1926–38). Other works include:

Viscount Astor and B. Seebohn Rowntree, *British Agriculture* (1938)

K. A. H. Murray, *Agriculture* (History of the Second World War) (1955)

P. Self and H. J. Storing, *The State and the Farmer* (1962)

Ministry of Agriculture, Fisheries and Food, *A Century of Agricultural Statistics* (1968)

CHAPTER 3

Land and Weather

Agriculture is not only an economic activity, but also a form of applied ecology. Crops and, to a lesser extent, livestock largely depend on the resources of their immediate environment, which can be modified only at some cost. To varying degrees, farmers also must accept the physical environment as given and select appropriate enterprises, and, at the national scale, there is increasing concentration of enterprises in those areas to which they are well suited. The marked differences in climate, relief and soils within Great Britain must therefore be a major factor in explaining the pattern of agricultural activity.

The reasons why types of farming vary throughout the country are exceedingly complex and there are few situations in which physical factors are either all-important or of no account. Much depends on the scale of investigation and on the enterprise studied. Over a small area inter-farm differences in management may be far more important in explaining agricultural variations than any differences in soil and climate, while Great Britain as a whole reveals regional contrasts which must largely reflect differences in environment. Similarly, physical factors are much less significant for the distribution of a crop such as oats, that can be successfully grown nearly everywhere in Great Britain, than for one such as maize, which is at its limit of cultivation. Unfortunately, neither the relationships between crops or livestock and environment, nor those between the economic and the ecological aspects of agriculture have been systematically investigated and the literature of both agriculture and geography contains little specific information about them. At this stage, therefore, the study of how physical factors affect the pattern of agricultural activity must largely be an account of the distribution of those factors which appear relevant, with tentative generalisations about probable relationships. Such is the purpose of this chapter, although detailed discussion of individual enterprises will be deferred.

Unlike the study of economic aspects of agriculture, which is severely handicapped by lack of data, there appears at first sight to be abundant information about the environment. In some senses, this impression is correct, for there are topographic, geological and often soil maps, besides numerous meteorological records; but the information is often not in suitable form for agricultural analysis or is not generalised at an appropriate scale. For example, many climatic data are available only as monthly or annual averages, although agricultural events rarely match these units of time; and while the primary data can sometimes be reworked into some more meaningful form, this is impracticable for the study of any large area. The level of generalisation also matters, for there is little value in trying to relate physical data for specific sites to highly generalised agricultural information, or vice versa. Thus agricultural data available only by counties or regions, such as crop yields, cannot be related to soils, nor can crop distributions mapped by fields be related to generalised meteorological data, such as potential evaporation. Physical factors have been stressed in geographical accounts of agriculture because the availability both of large-scale maps and of data for individual fields from land-use surveys enables relationships between the distribution of tillage or individual crops and, say, elevation, slope or aspect to be readily established, as in J. A. Taylor's study of the location in south Pembrokeshire of fields used for early potatoes. Such specific studies are few, and relationships have been even more rarely quantified; in any case, the approach is useful only for crops, for the distribution of grazing livestock changes daily and seasonally and may cover a range of environments.

Even where data are appropriate, their significance is not easily established, especially in Great Britain, where the environment is rarely sufficiently unfavourable to prohibit an enterprise altogether. As W. H. Hogg has pointed out, neither the optimum nor the limiting conditions for crops are known and, even if they were, matching environmental requirements at different stages in the life cycle is a task of extraordinary complexity. Nor does an empirical approach help, for, while certain minimum conditions must be satisfied before a crop can be grown or a class of livestock kept, the absence of an enterprise is rarely caused by the failure to meet these conditions; it normally results from an unacceptable return compared with that obtainable from other enterprises. In practice, therefore, we can rarely establish the environmental limits of an enterprise by observing where it occurs, because it loses its attraction to farmers long before these limits are reached. Thus, the present distribution of wheat does not indicate where wheat *can* be grown, as a comparison with its distribution during the Second World War or during the 1860s will demonstrate, and the question 'Where can wheat be grown in Britain?' invites the counter-questions 'At what yield per acre and with what degree of reliability?' The present distribution indicates the conditions in which farmers think it reasonable to grow wheat, given current prices for wheat and for alternative products, in other words, areas of comparative advantage for wheat. A further complication is that, where conditions for a particular crop are more restrictive, farmers appear to be more careful in choosing fields, so that, paradoxically, yields in counties where conditions appear to be unfavourable and only a small acreage is grown sometimes differ little from those in the main producing areas (although any extension of these small acreages would probably reduce average yields).

In a long-settled country such as Great Britain, there is no simple antithesis between the physical environment and the man-made environment. We now know that many present characteristics of land are products of past human activity and that, although the modification of natural conditions varies in degree, it is universal; the extreme contrast is between the watery wastes of the medieval Fenland and the intensively farmed first-class arable land of today, but even the open moorlands have been much affected by burning, grazing and other activities. We cannot easily quantify the human effort expended for generations in embanking rivers, digging drains and ditches, clearing stones and boulders from the land and modifying soil texture through claying, marling, warping and similar activities. It represents an immense investment, however, and it is continuing in such activities as draining land, applying lime and fertiliser and in other forms of land improvement. Without these changes, historical and contemporary, our scale of land values would be very different.

Climatic Factors

Physical factors affecting agriculture may be divided into climatic and physiographic, although these are clearly interrelated, with climate modified by relief and aspect, and soils by the balance of water supply and loss. The broad physical controls over the pattern of agricultural activity in Great Britain are climatic and their effects are greater in crop production than in livestock farming; for while livestock can be sheltered or moved when climatic conditions are adverse, crops cannot evade climatic hazards by movement or justify protection on a field scale; the subsequent discussion will consequently emphasise relationships with crop production, although this affects livestock through the provision of grazing and fodder.

The most characteristic feature of the climate of the British Isles is its variability, expressed in the quip that Britain has weather but no climate. Atmospheric conditions are dominated by eastward-moving cyclones, with their associated sequence of weather, and stable anticyclonic conditions obtain only occasionally. This variability is not confined to the day-to-day sequence of weather; it also characterises annual fluctuations and it is therefore debatable what climatic parameters are the most meaningful agriculturally. A farmer is likely to be guided in his choice of enterprises, not only by conditions in the previous season, however extreme they were, but also by his estimate of the likely sequence of weather based on past experience. Weather records in a particular year, e.g. the severe winter of 1946–7, may be quite misleading; equally, averages hide the range of variations from year to year, which may be vital in fruit growing, and, if taken over too long a period, may obscure secular changes of possible significance, especially on the margins of cultivation. There is no wholly satisfactory solution but, given that physical controls in Great Britain are generally indicative rather than absolute, the long-term averages provided by the Meteorological Office are a valuable guide.

Despite its small size, Great Britain exhibits a surprising range of climatic conditions. This arises partly from the disposition of high ground to the west and north, reinforcing the contrasts between east and west that result from the predominance of cyclonic weather and from the warm water of the North Atlantic Drift around the western shores, and partly from the maritime nature of the climate and the resulting steep lapse rate. Moderate changes in altitude produce marked changes in climate; for the growing season diminishes rapidly with increasing elevation, and the effects are accentuated by greater precipitation and windiness, especially on the western side of upland masses. Farmers can experience a range of climates within the confines of a single farm, especially in the west and north, and no small-scale maps can show such variations, which may be highly significant in the location of horticulture and other climatically-sensitive enterprises.

The major climatic determinants of plant growth, which vary widely throughout the country, are light, heat and moisture; they are also not unimportant to livestock, although some animal rhythms appear to be independent of climate. The variation in the availability of light can be predicted with certainty (although its intensity is affected by differences in cloudiness), while rainfall is the most variable of these climatic parameters. Day length varies with latitude and is some two hours longer in Shetland in mid-summer than on the south coast of England and correspondingly shorter in winter; these longer summer days largely compensate for the differences in intensity of insolation. Light controls the onset of certain biological processes, e.g. the breeding season in sheep or sexual maturity in poultry, and many plant species and varieties are adapted to the length of day; but light is not a major factor in accounting for the distribution of agricultural activity throughout Great Britain, although it is the one climatic parameter which glasshouse owners do not control. Sunshine decreases with increasing latitude and with altitude. Thus, average daily bright sunshine ranges from over $4\frac{1}{2}$ hours per day along the south coast of England to under $3\frac{1}{2}$ in north Scotland and over most uplands; summer sunshine similarly averages over 7 hours along the south coast and under 5 in Caithness. Sunshine is important in the ripening of fruit and in glasshouse cultivation (although air pollution may reduce the theoretical sunshine hours considerably in the lee of large cities) and late summer sunshine affects the sugar content of sugar beet.

TEMPERATURE

Temperature is less variable from year to year than rainfall, but in Great Britain quite small changes can be highly significant, especially in marginal areas. For the agricultural geographer, the two best indicators of regional differences in temperature currently available or easily derived are length of growing season and accumulated temperatures above the minimum for plant growth. The growing season is a highly generalised concept, for the temperature at which significant growth begins varies from species to species; conventionally 6° C. (42° F.) is accepted as the threshold temperature, although strictly it relates to the onset of grass growth. The growing season varies from over nine months around the coasts of the south-west peninsula to under seven in the uplands, but is between seven and eight months over most of the lowlands.

While the growing season imposes absolute limits on plant growth, most annual crops use only part of it; harvesting of cereals, for example, occurs up to three months before the growing season ends. Even with permanent grass unfavourable soil conditions, which prevent access to the land, or lack of moisture for grass growth may limit the effective grazing season. For many crops the onset of growing conditions is thus more important than the length of the growing season; on average, this occurs before 14th March in south-west England and coastal South Wales, between 28th March and 14th April in the main cropping areas and after 28th April around the Moray Firth (Figure 2). In more northerly districts, the onset of growth is often further delayed by low soil temperatures.

The length of the growing season is only a partial guide to suitability for crop production, because the rate of growth is affected by the range of temperatures above the minimum; although the relationship is not linear, the rate of growth increases with temperature, and certain minimum temperatures are necessary for the onset of the flowering phase. One measure of this relationship is the number of day-degrees, that is, the cumulative total of degrees by which the mean daily temperature exceeds the minimum for plant growth. This ranges from over 2,500 in southern England to under 1,500 over most of the uplands, with the lowlands of north-east Scotland, where

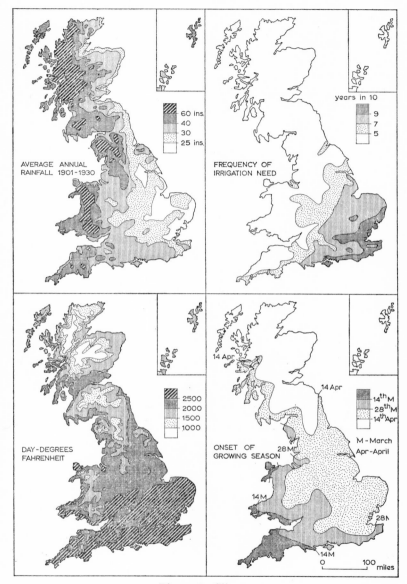

Figure 2. Climate
Based on maps and records of the Meteorological Office and on maps by the Ministry of Agriculture and by S. Gregory and J. A. Taylor

temperature is a major control, having between 1,500 and 2,000 day-degrees (Figure 2). As with length of growing season, not all this energy is used by annual plants, but the map does indicate where temperatures are most and least favourable for plant growth.

For livestock, temperature conditions control the length of the grazing season and provide one indicator of housing need. For crops, the critical consideration is whether temperatures are sufficient for successful ripening and harvesting. This is particularly true of cereals, for the drying power of the atmosphere decreases towards autumn and in September is only half that in August. A further complication is that late harvests

TABLE 1
SAMPLE HARVEST DATES

	1962	1963	1964	1965	1966	Average
Winter Wheat (Cappelle Desprez)						
Edinburgh	9/10	21/9	14/10	—	5/10	6/10
Isle of Ely	29/8	5/8	—	19/8	23/8	20/8
ALL PLOTS	5/9	22/8	9/9	27/8	28/8	30/8
Spring Barley (Proctor)						
Edinburgh	30/9	2/9	16/9	6/9	30/8	5/9
Hampshire	10/9	22/8	12/8	16/8	9/8	15/8
ALL PLOTS	9/9	19/8	31/8	25/8	24/8	30/8

Source: National Institute of Agricultural Botany.

may conflict with the demands of other farm work, such as autumn ploughing. A late onset of the growing season, combined with a small range of temperatures above the minimum, can thus severely restrict the crops that can be grown. Unfortunately, there is little information about harvesting dates, although test plots of the National Institute of Agricultural Botany show a general gradient from south and east towards north and west, where harvests on average may be a month later, and at least one large cereal grower transfers machinery from southern England to Scotland for the later harvest. There is, however, great variation from year to year, as is shown by Table 1, although the data must be treated with reserve.

An aspect of temperature and plant growth worthy of special mention is the incidence of frost. The frequency of frost increases inland and with altitude, but its occurrence is much affected by details of the topography, especially in relation to the free drainage of air, and risks can be minimised by careful siting; in central England, the frost-free period on a favourable site may be twice that on an unfavourable site in the same locality. With frost-sensitive plants, the date of the last frost is more significant than that of the onset of the growing season; for in most inland areas air frost is likely some time after 1st May. The incidence of frost is particularly important in the growing of fruit and some vegetables; for example, in 1953-4 frost and wind destroyed half the winter cauliflower crop in Cornwall and the 1967 fruit crop was well below average because of frost during the period of fruit setting. Frosts, however, are seldom sufficiently severe to damage winter vegetables, and the flavour of some crops, such as Brussels sprouts, can even be improved by autumn frost.

MOISTURE

Data on the availability of moisture are more meaningful and are also more soundly based, since far more stations record rainfall than record temperature. Annual totals of rainfall show the marked contrasts resulting from the prevalence of easterly-moving depressions and the concentration of high ground in western counties, for areas with an annual rainfall of over 60″ broadly correspond with land over 1,000′ (Figure 2). Precipitation increases with altitude but there is also a west-east contrast on the larger uplands, particularly in Scotland, where the eastern Grampians receive less than 60″, compared with more than 100″ on the mountains farther west. Within the lowlands there is also a gentle west-east gradient, ranging from over 40″ along the west coast to under 25″ in parts of East Anglia, although favoured localities such as lowland Shropshire, in the rain shadow of the uplands, have a lower rainfall than some areas farther east, and higher tracts, such as the Chilterns, receive several inches more than the surrounding lowlands. The pattern of rain-days (i.e. days on which more than 0·1″ of rain falls) is broadly similar, ranging from over 250 in the north-west to under 175 in the south-east. All these figures are, however, averages, and precipitation in any year may not only differ by as much as 50 per cent. from these values, but show opposite trends in different parts of the country, as in the summer of 1968, when the weather in Scotland was dry and fine and that in southern England was cloudy and wet; furthermore, the variability is generally greatest in the areas of low precipitation in the east. Regional contrasts in summer rainfall are less, for whereas the Scottish Highlands have a winter maximum, East Anglia has a summer maximum, over most of the rest of the country the second half of the year is generally wetter.

More important for plant growth than total precipitation is the balance between precipitation and potential evaporation. While in most of the country potential evaporation exceeds precipitation in one or more summer months, there is a contrast between the north-west, where soils rarely dry out and less radiant energy is available for heating, and the south-east, where summer usually has a negative water balance, with soils drying out and plants receiving less moisture than is needed for maximum growth. This situation is better than excessive moisture in the soil, for while water can always be added (at a cost) drainage is not possible everywhere. On average the soil moisture deficit exceeds 6" around the Thames estuary and 3" over most of East Anglia, and it appears that irrigation water would be useful on grassland (which suffers most from this deficit) nine years in ten on land bordering the Thames estuary and less frequently in other parts of southern and eastern England (Figure 2). Whether such water is applied is a matter of availability and of the relationship between the cost of applying water and the increased returns that result.

The consequences of differences in precipitation are, like those of temperature, complex and difficult to isolate. Thus the effects of precipitation deficit on annual crops are less than might be expected, since much of the growth is made when the deficit is small; for example, as much as three-quarters of the growth of cereals is probably made in the first half of summer. High rainfall is probably detrimental to all kinds of agricultural production, although in part indirectly, through the acidity caused by heavy leaching and through excessive soil moisture. Thus, it may be difficult both to get on to land and to work it, early or late grazing may be lost, the sowing of crops delayed and harvesting handicapped; heavy rainfall may also adversely affect the quality of hay and depress cereal yields. Conversely, grass growth is reduced by shortage of moisture, although the late summer decline in growth is only partly because of this. There is thus a sound meteorological basis for the concentration of grassland in western lowlands and of arable crops in the east; L. P. Smith has shown that yields of meadow hay correlate well with estimated potential transpiration, and that the pattern of grassland is closely associated with the distribution of 'favourable' climate. Conversely, cereals, being deeper rooting, are better suited than grass to areas with a marked soil moisture deficit, where the range of possible crops is much larger.

Rainfall is the ultimate source of both surface and underground water, and might be expected to affect the distribution of livestock; for cattle require about 10 gallons a day and sheep about 1½ gallons (even if half the former and all the latter is normally provided from herbage). Lack of drinking water for livestock has certainly inhibited the keeping of cattle in parts of eastern England in the past, but with more than two thirds of farms in England and Wales connected to mains supplies, water is no longer important.

Other climatic factors reinforce these contrasts. Wind speeds are highest along the west coast and increase with altitude, their effect depending greatly on exposure: driving rain is much more frequent in such localities. Cloudiness and humidity also increase with altitude, as does snow cover, which averages over twenty days a year on much of the uplands and under five in the south-west peninsula. All these conditions contribute to the unsuitability of the uplands for most forms of agriculture and to the low returns obtainable from the few possible enterprises.

Late frost is one exceptional meteorological hazard in agriculture which has localised consequences. Hail is another, and the highest incidence recorded in the period 1933–47 occurred in the premier fruit-growing county, Kent, which averaged more than twenty-five days, and the least in Leicestershire, which averaged less than one. Heavy thunderstorms can likewise cause severe damage to crops and are more frequent in the main arable areas in east and south-east England. Such climatic hazards are increasingly the subject of insurance policies.

Many effects of climate are felt incidentally through the provision of feed for livestock or through the incidence of disease. With few exceptions, such as potato blight, where the risk of severe attack can be forecast and prophylactic action taken, the relationship between meteorological conditions and outbreaks of disease is still imperfectly understood. The area subject to air pollution, which adversely affects both crop production and animal health, is also largely determined by meteorological conditions, being generally more extensive to the east of towns.

The nature of both atmospheric conditions and agricultural production suggests that their interactions are highly complex. While there is often good general agreement between the present pattern of agriculture and what might be predicted on meteorological grounds, it is impossible to obtain optimum conditions for all forms of agricultural production or for all stages of any particular enterprise. Different requirements may conflict: thus yields of potatoes and sugar beet are increased by early planting, but this also increases the risk of damage by late frost. Furthermore, the generality of conditions matters less than those occurring at particular stages in development: thus, while hardy sheep and cattle can withstand severe climatic conditions, weather during the few days immediately after birth can be critical.

The effects of meteorological conditions can be moderated in various ways. Animals can be moved to more favourable environments, as with the wintering on lowland farms of ewe lambs from hill flocks. The provision of shelter belts or hedges can also benefit both livestock and crops, especially early horticultural produce. Local climate can, in effect, also be modified by improving soil drainage (i.e. making the local climate drier) or by irrigation, as in many vegetable-growing areas. The area irrigated rose sharply from the early 1950s to the early 1960s, and totalled some 250,000 acres in 1964, less than 3 per cent. of the total area under crops and only a sixth of the area which might be irrigated; further development has been inhibited by increased charges for water. Control becomes more complete when crops are grown and livestock kept under cover. Such protection may be only periodic, as with the farrowing of pigs, or seasonal, as with the housing of cattle, with animals kept in the open in summer and housed during the winter. The period for which cattle are housed is at least partly related to climatic conditions and, according to a sample survey in the 1950s, cattle are never housed by day in the south-west peninsula, but are housed day and night for six months in Scotland and northern England. The chief value of such shelter is to limit exposure rather than to provide warmth. Growing crops under cloches and cold frames represents an intermediate stage, where temperature conditions are modified, but heated glasshouses provide almost wholly artificial conditions, with increasing use of automatic control of temperature and shade and even a modified atmosphere enriched with carbon dioxide. Even in glasshouse production, the cost of heating is itself related to climatic conditions, and an additional ton of coal is estimated to be required to heat an acre of glass for every day-degree drop in temperature. The tendency for farming to go under cover has gone furthest in intensive livestock rearing, especially poultry, which are kept throughout their short lives in wholly artificial environments with temperature and light completely controlled.

Evidence of the effects of climate on agricultural production is scanty, but climate is patently a major consideration in both the choice of individual agricultural enterprises and the range of enterprises possible in different parts of the country; for climatic limitations can be overcome only by careful management, by expenditure of capital or by continuing outlays on other inputs. The variability of climate is a cross the farmer has to bear and he must himself decide on the probability of favourable weather conditions and on the likelihood of profitable production. At the present stage of knowledge, what farmers in fact do is probably as good a guide to relative suitability as any consideration of climatic data.

Physiographic Factors

The part played by the land itself is little better known than that of climate, but its influence is more likely to be felt at the

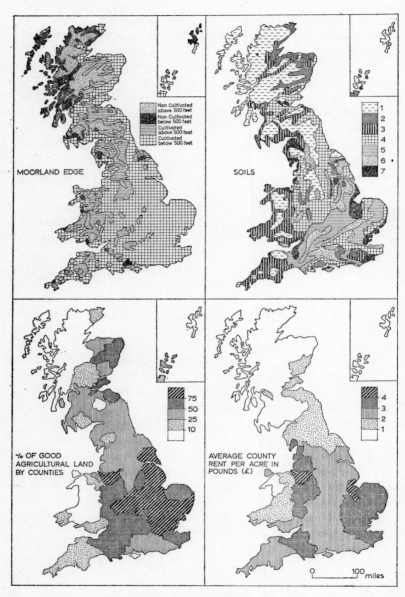

Figure 3. Land and land quality

level of individual farms and even fields. It can be conveniently discussed under two heads, relief and soils.

RELIEF

The influence of relief has already been noted during the consideration of climatic factors, in its role of modifying climate. Mean temperature falls by about 1° F. for every 300' of elevation and G. M. Manley estimates that the growing season is reduced by approximately ten days for every 260'; it is therefore halved at about 2,000' and crops can seldom ripen above about 1,200'. The increase in precipitation, cloudiness and humidity, and wind speed, already noted, and a decrease in day-time temperatures are further handicaps to agricultural production, and the range of crops diminishes and yields fall with increasing altitude. R. E. Hughes has shown that productivity of semi-natural vegetation in Snowdonia was halved between, 1,200' and 3,000', and R. W. Philips instances marked falls in crop yields in a transect from lowland Pembrokeshire to upland Cardiganshire. For livestock, too, conditions deteriorate with height and natural shelter becomes an important consideration in keeping even hardy breeds of sheep, although the main effect is felt through the availability of grazing, which may be inadequate even to maintain body weight. These handicaps find expression in the location of the moorland edge, the upper limit of cultivation, although it is not always easily recognised on the ground; for it is the highly complex product of many forces, physical, historical, economic and personal. If, however, it is generalised, e.g. by computing the average maximum elevation of continuous improved land, and examined at a small scale, certain characteristics are recognisable (Figure 3). It is lower in the north than in the

---

Figure 3.
Based on maps of the Land Utilisation Survey, on data provided by the agricultural departments and on a map by D. A. Osmond. The county rent map is based on a complete enumeration in 1960 in Scotland, but on estimates from a sample survey of England and Wales in 1966.
The key to the soil map is: 1, podsolised and organic soils of the uplands; 2, podsolised soils of the lowlands; 3, acid brown soils; 4, grey-brown podsolic soils; 5, grey-brown podsolic and brown forest soils; 6, brown forest soils with rendzinas; 7, soils on alluvium and organic soils of the lowlands.

south, and in the west than in the east, reaching sea-level in north Scotland and attaining heights of over 1,400' on the Welsh borderland and even higher in the Pennines. It is also lower on small upland masses than on large. In detail, it is much influenced by slope and aspect, being lower where slopes are steep and aspect unfavourable. The moorland edge is not the upper limit of land which *can* be cultivated; as with crop distributions, it is the resultant of present technological and economic conditions and, as abandoned farmsteads and fields indicate, it has been considerably higher in the past, when levels of living were lower and self-sufficiency greater.

The slope of the land is also a factor, although data about slope exist only in generalised form as contours on maps and there is little evidence from which its significance can be evaluated. One would expect the use of machinery to become more difficult with increasing steepness of slope, but there is little information about how cultivation costs vary with steepness, nor can any very meaningful practical limits of safe operation for different kinds of equipment be specified: much depends both on present economic circumstances and on the way in which equipment is operated e.g., one-way ploughing with a crawler tractor is possible on virtually any slope with a covering of soil. W. J. West's estimates of limits of slopes for normal working of various implements range from $20°$ for a track-laying tractor, through $15°$ for a wheeled tractor, to $4°$ for a combine harvester. Despite this lack of evidence, operating costs probably rise with steepness and farmers will not cultivate steeper slopes where alternatives are possible. Slope may also affect livestock farming, for movement on steep slopes involves the consumption of additional energy by animals, and the consequent loss of production is a disincentive towards the keeping of dairy cattle in such conditions.

Within the lowlands, slopes are rarely steep enough to be prohibitive and many of the scarp slopes, which are generally uncultivated, could be ploughed if necessity demanded. Mechanised cultivation is facilitated by large tracts of level or near-level land, provided these are well-drained, and the main arable areas satisfy these conditions. In the west and north, on the other hand, stretches of gently-sloping land suitable for cultivation are confined to the coastal lowlands. Within the uplands there are three major slope components, each, for various reasons, little suited to cultivation. The slopes of incised valleys are often too steep to warrant ploughing and, paradoxically, the valley bottoms are often marginal because of a high water-table; on the gentle slopes of the uplands, both poor drainage and unfavourable climatic conditions limit agriculture.

Poor drainage may thus be caused by topographic conditions, notably in areas with a high water-table, although most areas with impeded drainage are low-lying. Much of this land has been banked and drained, and parts of the Fenland, the largest such area in Great Britain, lie below sea level, the water-table being controlled by pumping. While artificial drainage works are properly maintained, such land is among the most fertile in the country; but low-lying land can also be marginal land, and W. Ellison estimated in 1953 that there were some 800,000 acres of such poor land. In such conditions, access to land is restricted and yields may be low; in the National Wheat Survey, poor drainage appeared as the major soil factor affecting wheat yields, which averaged 20 per cent. higher on moderately drained than on poorly drained soils. Over most of the lowlands, natural drainage has been much influenced by the embankment of rivers and coastal flats, and by the under-draining of large areas. How much land is under-drained is not known, for under-draining is not permanent; but it is estimated that 13–14 million acres in England and Wales could be improved in this way and that drainage would be economically justifiable on half this area. Much low-lying land is also liable to flood when high tides pond back rivers or breach dykes, as in the East Anglian floods in 1953, or when heavy rains or melting snow on already saturated soils produce rapid increases in runoff and river discharge. Engineering works necessary to minimise such damage by floods are expensive; but, generally, floods are occasional rather than frequent and the area where the flood hazard seriously impedes agriculture is probably quite small.

The influence of relief is also expressed in aspect, although here, too, its effects are indirect through the modification of climate and are primarily local; there may be a month's difference between a south-facing and a north-facing slope, although exposure to wind may moderate or accentuate the effect of differences in insolation. Aspect is important in growing early crops, for small differences in earliness greatly affect prices, but variations in aspect within the same farm may also extend the grazing season by providing both early and late pastures.

Most areas too rugged to be cultivated are, in any case, climatically unsuited to cultivation because of their elevation; but, even in the uplands, rugged country is largely confined to Snowdonia, the Lake District and the Scottish Highlands, and large areas of upland are capable of being ploughed. At lower elevations there are limited tracts of limestone pavement and of thin soils with limestone near the surface, as in the Mendips and the Peak District, where cultivation is either handicapped or prevented altogether. Where glacial deposits have been little altered by erosion, as in parts of the Scottish Lowlands, relief is often very broken, with ill-drained hollows, steep slopes and boulders, and these conditions also handicap mechanised cultivation. Apart from the features resulting from the concentration of high ground in the west and north, the direct influence of topography on agricultural activity is thus felt chiefly in respect of local details.

SOILS

More has probably been done to remove the handicaps imposed by soils than to overcome any other physical difficulties, and most soils have been improved from their natural state to become, in effect, agricultural soils; but documentation of the relationships between soil and agricultural production is also generally lacking.

In detail, the pattern of soils in Great Britain is complex, reflecting the variety of parent materials (especially in glaciated areas), the range of climates and the small-scale topography, and soils may vary both within farms and even within fields. Most of the country has yet to be mapped in detail so that one must often rely on maps of drift geology or on small-scale reconnaissance maps. In outline, soils fall broadly into four groups (Figure 3). The uplands, where climate is generally inimical to cultivation, are characterised by podsols, peats and peaty podsols, all of low fertility and often with impeded drainage. The second group consists mainly of the acid brown soils of the upland margins, with below-average fertility and low base status. The third group, the grey-brown podsolic soils, occupying mainly lowland sites, are not dissimilar, but have greater potential because of their more favourable climatic environment; in low-lying areas, however, they often have impeded drainage. Lastly there is a group of soils, consisting mainly of brown earths, but including calcareous soils and those developed on alluvium and on fen peat, which are adequately but not excessively drained and have a relatively high base status; they provide the better agricultural soils and are generally ploughed. Fertility as such now exerts less influence in the lowlands than formerly because deficiencies can be remedied by artificial fertilizers, and soil drainage and workability, which are less easily modified, are now more significant factors. Consequently, some light, well-drained soils have come to be highly regarded, especially where they occur on large areas of gently sloping land which favour mechanised cultivation, although they are less suitable for grass and other shallow rooted crops in areas with a marked soil moisture deficit. Returns from upland soils do not generally justify heavy investment in drainage or fertilisers.

The agricultural significance of soils depends on climatic conditions. In the areas of heavy rainfall, loss of nutrients through leaching can be offset only by heavy inputs of lime and fertiliser, which place a considerable strain on farmers' resources. Heavy textured soils are slow to warm up in spring and unusable by either machines or grazing cattle for long periods in the winter, especially between November and February. The risk of damaging grassland by treading such soils thus restricts the early and late grazing and limits the length of the effective grazing season, while ploughing may have to be delayed till spring. Costs of cultivation also appear to be higher on heavy land. For example, I. Reid gives machinery costs for barley growing in south-east England in 1961 as £7·4 per acre for heavy land, £6·9 for medium land and £6·5 for light land (although these differences represent only 5 per cent. of variable costs). Additionally, heavy soils can be cultivated on fewer days and are more limited in the range of

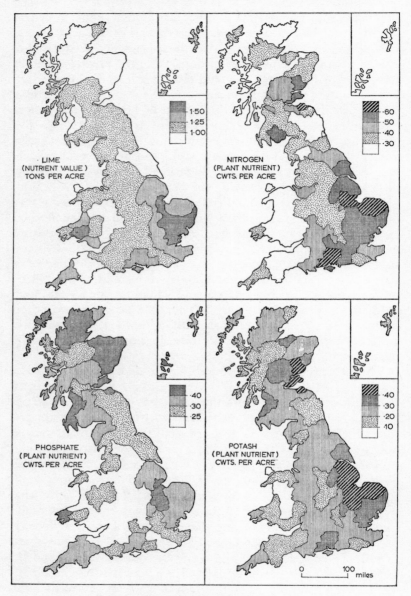

Figure 4. Lime and fertilisers
Source: Fertilisers Scheme, 1965

crops that can be grown successfully on them. The choice of crops is also limited on shallow and stony soils, like those derived from the Chalk and clay-with-flints respectively. As has been indicated, it is soil drainage rather than texture which is most important for wheat yields, although the evidence is insufficient for dogmatic statements about crops in general, especially as differences in previous cropping, fertiliser practice and management can be such important factors in yields. For most purposes, however, soils which are neither excessively nor poorly drained and can be worked most of the year are the more satisfactory for arable cultivation, while soils of heavier texture are more suitable for grass.

FERTILISERS

The nutrient status of the soil is the characteristic most easily altered. Soils can no longer be considered without reference to the inputs of lime and fertilisers, which totalled £90 million in 1966 net of subsidy or 10 per cent. of all farm expenditure, although the effects of these improvements in soil status are short-lived, especially in western districts, where leaching may be five times as severe as in eastern counties. Over 6 million tons of lime are now applied to the soil annually, more than sufficient to replace losses by leaching, and $4\frac{1}{2}$ million tons of fertiliser (or, in nutrient equivalent, 590,000 tons of N, 435,000 tons of P_2O_5 and 382,000 tons of K_2O). Applications on this scale are quite recent and only some half a million tons a year of lime were being applied in the whole of the United Kingdom before the introduction of the lime subsidy in 1937. Pre-war consumption of fertiliser was also much lower, being respectively some 60,000, 170,000 and 75,000 tons (nutrient content).

Applications of lime and fertilisers are far from uniform throughout the country and are generally highest in eastern counties (Figure 4). These maps are, however, somewhat misleading since they show average dressings per acre of crops and grass; thus, the acreage of improved land in the Highland counties is small and the actual quantities applied per acre of agricultural land (i.e. including rough grazing) is very small. The chief reason for the east-west gradients revealed by these maps is the lower rate of application to permanent grassland, which predominates in northern and western counties, and the smaller proportion of the acreage under crops and grass receiving dressings of fertiliser; the rate of application to those fields which are fertilised does not differ greatly. Perhaps surprisingly, the proportion of total inputs provided by fertilisers in the Farm Management Survey sample varies only from 6 per cent on pig and poultry and horticultural holdings to 14 per cent. on arable (mostly cereal) holdings, although the range in actual quantities per farm is very large.

Land Classification

Since the various physical controls over agriculture act together, not in isolation, cannot they all be subsumed into some measure of land quality? There have been attempts to do this. Thus, in 1939 M. G. Kendall constructed an index of crop productivity in England, using county yields for ten widely-grown crops (wheat, barley, oats, potatoes, turnips and swedes, mangolds, beans, peas, meadow hay and seeds hay), although he emphasised that the differences recorded would partly be due to factors other than natural conditions. Nearly all his correlations between yields were positive, indicating that areas where yields were good for one crop were likely to have good yields for other crops. His rating of counties, which placed Holland and the Isle of Ely in the top category of excellent and most western counties in the lower categories, accords broadly with what a consideration of physical criteria alone would suggest; but unexpectedly low ranking in some counties, such as Leicester, where there was little land in tillage crops, suggested that this classification undervalued suitability for grassland.

At about the same time, L. D. Stamp was preparing a classification of land based on both the physical qualities of the land and on the past and present use, and a map showing the distribution of the different classes of land was published at a scale of 1 : 625,000 in the Ordnance Survey's National Atlas Series. This map was not the product of a detailed field survey, but a general appraisal based on various sources, and large tracts of land had to be shown as occupied by a mixture of categories. It was not intended as a basis for individual decisions on land use. Three major categories of agricultural land were recognised, good, medium and poor, and these were further sub-divided, according to their suitability for arable, grass and mixed farming, to give a total of ten categories. The criteria used in classifying were qualitative, based on subjective judgments. Thus, first-class land was land that was highly productive under good management, being level or gently undulating, not too elevated and with a favourable aspect and deep, well-drained soils. Medium quality land was only moderately productive, even under good management, because of the operation of one or more factors of site, such as high elevation, steep slopes, unfavourable aspect and shallow or poorly drained soils. Poor land was of low productivity through the extreme operation of one or more factors of site or soil, such as wetness or elevation.

TABLE 2
LAND QUALITY IN GREAT BRITAIN

	acres '000	%
First class land	2,346	5·1
Good land	11,630	25·5
Medium quality light land	2,700	5·9
Medium quality other land	12,111	26·3
Poor heavy land	880	1·9
Other poor land	16,519	36·9

Source: Land Utilisation Survey.

Table 2 gives the acreages in Great Britain under the main categories, although some non-agricultural land is included; it illustrates the limited amounts of good land and should be read with Figure 3, which shows the percentage of such land in each county. The important feature is the location of most of the poor land in the high rainfall areas of the west and north and most of the first-class and good land in eastern districts.

The classification is still the only one available for the whole of Great Britain, but it has been criticised on several counts. At least explicitly, it does not take account of climatic differences, although these can lead to otherwise similar land being differently graded; for example, terrain which suffers from excessive drainage under light rainfall may deserve higher grading in western districts of heavier precipitation. Ideally the map of land classification should be read alongside a climatic map which would indicate the climatic constraints on agricultural production. The classification has also been criticised for overemphasising the farming practices and land use of the 1930s, but no classification can be permanent or even semi-permanent, given the rate of scientific progress and technological change. Post-war developments in light-land arable farming, especially on the chalk downs, make it necessary to reappraise such areas and other revaluations will undoubtedly occur, although no comparable revolution in heavy-land farming seems in sight despite greatly increased mechanisation.

Work is in progress on new land classification maps, at larger scales and based on field survey. In England and Wales the Agricultural Land Service is completing a survey at a scale of 1 : 63,360, in which five grades of agricultural land are recognised, the middle grade representing average conditions on a national scale. This classification, too, is essentially a subjective physical assessment, although specific values provide guidelines for allocating land to particular classes. Unlike the Land Utilisation Survey's classification, climate is explicitly included as one of the variables, but the gradings of the different classes are similarly based upon limitations of agricultural use. In Scotland, surveys of lowland areas at the 6″ scale have been made by the staff of the Department of Agriculture and Fisheries for Scotland, using the field as the unit of classification; three major grades of A, B and C quality land are recognised, and these are further sub-divided into six categories. A classification similar to that of the Agricultural Land Service is now being made at a scale of 1:63,360.

Land classification maps are useful as broad indicators of the suitability of different areas for agriculture. Ideally, there should be a separate evaluation for each enterprise, with one classification for wheat, another for potatoes and so on, for the requirements of different crops vary. Such a survey of land suitable for horticulture in England and Wales has in fact been made by the Agricultural Land Service, but surveys of suitability for individual crops seem hardly practicable given available resources. In any case, the quality of land as a physical entity (including climate) is only a partial measure of suitability for agriculture; for bare unequipped land is virtually unknown.

FURTHER READING

J. S. Bibby and D. Mackney, *Land-use Capability Classification*, Technical Monograph No. 1, The Soil Survey (1969)

N. Hilton, 'Land classification', in *Studies in Applied Geography*, Spec. Pub. No. 1, Inst. Brit. Geogr. (1968)

G. M. Howe, 'Climate in relation to crop production, *Agric. Progress* 32 (1957)

R. J. C. Munton and J. M. Norris, 'An analysis of farm organisation', *Geogr. Ann.*, Series B, 52 (1969)

L. P. Smith, *Farming Weather* (1958)

L. D. Stamp, *The Land of Britain: its Use and Misuse* (3rd Ed. 1962)

J. A. Taylor (ed.), *Weather and Agriculture* (1967)

Ministry of Agriculture, Fisheries and Food, *The Farmer and his Weather*, Bull. No. 165 (1966)

Ibid, (Agricultural Land Service) *The Classification of Agricultural Land in Britain*, Tech. Rept. No. 8 (1962)

CHAPTER 4
Farms and Fields

To the economist the term 'land' includes both the physical qualities of soil and climate and the fixed equipment which is the product of past investment. If man-made modifications make it inappropriate to think in terms of the 'original and indestructible powers of the soil', it is equally misleading to consider land as bare and unequipped. In Great Britain such land does not exist, nor in practice can one separate land in the physical sense from the structures erected upon it. For the land is divided into farms, which are the units of decision-making in agriculture, and these are sub-divided into fields by fences, hedges or walls, equipped with buildings and crossed by roads and tracks by which men, machines and animals circulate to make the farm a functioning unit. While each farm is unique and every county contains a mosaic of farms of different shapes and sizes, there are nevertheless regional characteristics which are of some importance in accounting for the regionalisation of farming types. Often these man-made features reinforce the influence of land in the more restricted sense, with land of good physical quality occupied by well laid-out farms, as in much of East Anglia, and the handicaps of poor land aggravated by poor farm layout, as in parts of the Midland clay vales. Sometimes, however, they work in the opposite direction, advantages of good physical conditions being reduced by poor farming structure, as in parts of south-west England, and the handicaps of poor land by good structure, as in the Southern Uplands. These man-made features represent a large investment of human and material resources over many generations and are not always appropriate to present needs: as a Minister of Agriculture has said, 'British farmers are attempting twentieth century farming with nineteenth, eighteenth or even seventeenth century fixed equipment'. This chapter is concerned with such characteristics, with the changes that are affecting them and with their significance for agricultural production.

THE SIZE OF FARMS

Although we know that the land is divided into farms, this is not immediately observable. From the air, the agricultural landscape appears to be divided into fields rather than farms, whose boundaries are not distinguishable from field boundaries marking internal divisions of farms. It is true that the tenanted farms of a large estate may have common characteristics that distinguish them from other farms and an individual owner-occupier may adopt some distinctive material to mark the perimeter of his farm; but these are exceptions. The distribution and character of farm buildings provide some indication of farm size, but they must be used with caution. In areas of stable farm structure and dispersed settlement, the spacing of farmsteads generally provides a good index, but it is more difficult to assess where farmsteads are located in large villages, especially where holdings are small and their associated farmsteads not easily distinguished from other buildings. The widespread trend towards farm enlargement is an additional complication, for many redundant farmsteads are used to house agricultural workers or sold as private residences or holiday homes, and not all the buildings named as farms on Ordnance Survey maps are still farms. The size of the farm buildings offers some guide, but much depends on type of farm; the development of intensive farming systems requiring large buildings and little land presents further problems.

Such problems could be easily overcome by consulting maps showing farm boundaries, but Great Britain is almost unique among developed countries in not having comprehensive public records of the ownership and occupation of land. The surveys made in England and Wales in the 1830s and 1840s for the commutation of tithes, which cover some three quarters of these countries, provide the information for the construction of maps of farm and estate boundaries of that period. More recent farm boundary maps have been prepared by the agricultural departments, but are not generally available. In England and Wales maps were constructed for the National Farm Survey between 1941 and 1943, but they are now out-dated and were often incomplete in areas of complex land holdings. Farm boundary maps of Scotland have also been prepared since the Second World War by officers of the Department of Agriculture and Fisheries, but these, too, are now some twenty years old.

Even if accurate contemporary maps were generally available, they would present considerable practical problems for the study of a large area, so the information derived from agricultural returns must be used for any country-wide survey. It is important to appreciate that this information relates not to farms, but to agricultural holdings i.e. parcels of land exceeding one acre and used for agricultural purposes. Most of the large agricultural holdings are indeed farms in the generally accepted sense, but many small holdings are not, and include accommodation fields, the grounds of large institutions, playing fields and golf courses where these are grazed; such semi-agricultural uses are particularly common around large cities. Many others are run as spare-time holdings by those with full-time occupations elsewhere and others are part-time holdings (see Chapter 5). While many of the latter can be regarded as farms, most of the remainder cannot. It is therefore misleading to determine average farm size by dividing the area of agricultural land by the number of holdings.

A further complication, of growing importance, is the development of multiple holdings, that is, several holdings which are farmed as a single unit. The holdings may not be physically contiguous, since adjacent farms become available for purchase or letting only infrequently and fragmented multiple holdings may represent only intermediate stages in the development of large compact holdings. There is, however, a general impression (though no firm statistical evidence) that it is increasingly common for widely separated holdings to be linked in this way under the same owner or tenant and operated under a common plan; for example, a holding on the uplands may supply another in the lowlands with young stock for fattening and receive fodder crops in return. It is, of course, debatable how wide separation must be before such holdings cease to constitute a single farm, although resources can be moved over surprising distances; a farmer in south-east Scotland with land in the English Midlands regularly exchanges harvest equipment between his holdings. Where there is regular movement of labour, machinery and stock between different holdings and where they are managed as a single unit, they may properly be regarded as a single farm; but where they are under different, if related, managements and there is only occasional movement of resources between farms, it is more appropriate to treat them as separate farms. The Ministry of Agriculture's ruling is that where holdings are farmed together they should be returned as a single unit, provided they lie within the same county; but holdings have been known to be returned separately after amalgamation and a changed wording of the census instructions in Scotland in 1957 was followed by a 7 per cent. drop in the number of holdings returned. Sometimes there are reasons, such as the payment of subsidies, where a farmer profits by returning holdings separately; but the practice is often the product of inertia and even large estates often continue to refer to amalgamated holdings under the names of their component farms. Whatever the reason, it is highly likely that data on agricultural holdings understate the number and extent of large farms and exaggerate the number of small holdings.

Further difficulties arise in examining changes in the size of farms over a period, especially where the comparison is confined (as it generally is in any historical study) to numbers of holdings in different size groups rather than to the acreages occupied by them; for the numbers are not strictly comparable. When the first agricultural returns were collected, many holdings were unknown to the officials responsible and were omitted, but the returns became progressively more complete, although as late as 1936 E. Thomas and C. E. Elms found 349 holdings, including some substantial farms, in Buckinghamshire which were not being returned, and the introduction of livestock rationing in 1941 revealed some 270,000 acres which had escaped enumeration. Changes in administrative practice also make comparison difficult. In the nineteenth century farmers with more than one holding were given latitude regarding the number of returns they made, but rules have since been tightened and in 1956 a special scrutiny of the lists of holdings in England and Wales led to a reduction of some 8,000 in the total number of holdings, partly by eliminating units which should not have been included in the returns, partly by the amalgamation of once-separate holdings. Where the allocation of holdings to size groups is made by acreage in crops and grass, other illusory changes may result from the reclassification of land under rough grazing as grassland and vice versa. Where run-down permanent grassland is subsequently returned as rough grazing, as commonly happened between 1880 and 1939, the size of holdings appears to fall even though their actual boundaries remain unchanged. Other difficulties arise over land which is, or is believed to be, common land, and so should not be included in the returns of occupiers with rights of common grazing; for example, in 1938 it was found that many occupiers in Breconshire were returning common land as part of their own holdings. Improvements and changes in the machinery of enumeration and the reclassification of rough grazing have contributed to the apparent rise in the number of small holdings and the decline in the number of large early in this century which have been noted by several writers.

THE DISTRIBUTION OF FARMS

Nevertheless, data from the agricultural returns on size of holdings provide the most useful and convenient source for the study of regional variations in farm size, but somewhat different pictures emerge if the grouping of holdings is made by area of crops and grass instead of crops, grass and rough grazing (or in Scotland, total area, i.e. including steadings, roads and farm woodlands). If size is measured by acreage of crops and grass, the main arable counties stand out as areas of large farms, with more than a quarter of their improved farmland in holdings of 500 acres and over, the chief exceptions being the Fenland, with many small holdings, and Essex, which includes part of the London fringe (Figure 5). Nearly all the upland counties have over a quarter of their crops and grass acreage in holdings

Figure 5.
Source: agricultural censuses 1965. The upper maps show the area (of owner occupied land or common grazing) in each county as a proportion of the total area of agricultural land. The lower maps show the proportion of the area of agricultural land and crops and grass respectively in large holdings(A), i.e., of 500 acres and over, and small holdings(B), i.e., under 100 acres, the size of holding being measured by the acreage of agricultural land (left hand) or crops and grass (right hand) on each holding.

FARMS AND FIELDS 59

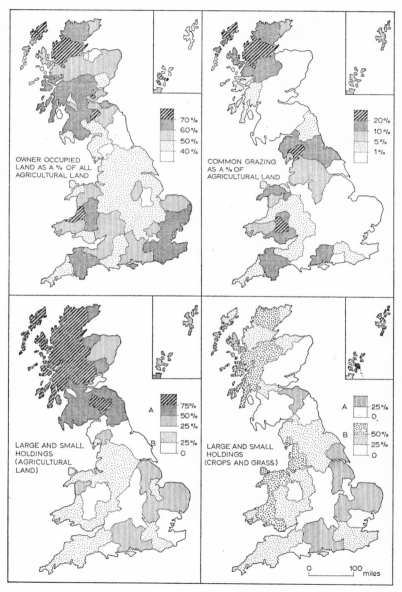

Figure 5. Holdings and tenure

of under 100 acres, and eleven counties have over half. If size is measured by total farm area or by the acreage under crops, grass and rough grazing, interesting contrasts emerge. In lowland England the changed basis of classification has little effect, but in Scotland large holdings now occupy over half the farmland nearly everywhere and more than three quarters in the Highland counties. In Wales, and to a lesser extent the English parts of Highland Britain, the differences between the two classifications are less and holdings in these counties can be regarded as small on either basis.

The reasons for the regional differences are numerous. On general economic grounds, farm size might be expected to vary with land quality, so that the poorer the land, the larger would be the area of the farm. Such relationships can be seen on these maps, for the largest farms, those on the hill lands of north Scotland, lie on the poorest land, while the best land, the Fenland, has a high proportion of small holdings. More generally, such a relationship is expressed through the association between farm size and type of farm; thus arable farms are larger on average than dairy farms and hill sheep farms than livestock rearing farms. The relationship is only a broad one, for a wide range of sizes is found among farms of any one type and within any particular area, but there are considerable regional differences. Within the lowlands, small holdings characterise the fringes of large cities, often providing spare or part-time employment in agriculture for those with other sources of income; many of these are agricultural holdings only by courtesy. Other anomalies reflect historical circumstances. The most obvious of these (although not evident on a county map) is the concentration of small holdings around the coasts and on the islands of north-west Scotland; in the seven crofting counties, statutory security of tenure permits the survival of many tiny crofts which, on economic criteria, would long since have disappeared. There are many other examples, and the contrasts between Scotland and Wales are partly a product of such historical differences. In Wales many small farms were established by squatters, who illegally enclosed parts of upland commons, and others as part-time holdings by miners, as in areas of lead mining in Cardiganshire and in the Carmarthenshire part of the South Wales coalfield. In the Scottish

Highlands, on the other hand, the land was cleared for large sheep farms and the small holders were often resettled on the coasts or migrated overseas. The fact that farming is only in part an economic activity, so that those who prefer independence to high wages may continue to farm against all economic logic and that changes in farming structure in a free society are effected only gradually also contributes to the lack of close correspondence between farm size and land quality.

There is currently a general tendency for both the average size of farms and for the proportion of farmland occupied by large farms to increase. In part this trend reflects rising standards of living, which imply an increase in the size of farm necessary to provide an acceptable level of living (at least for enterprises which cannot be intensified), and also rural depopulation, which may lead to the abandonment of farmsteads when the occupier dies or retires and the incorporation of the land in another farm. But in recent years it has been primarily due to economic forces, especially to pressures on farmers' profit margins, which make it necessary to maximise the use of resources of labour, fixed equipment and machinery. In one study in Yorkshire, where 5,417 acres were added to existing holdings, only 13 extra men were required. The trend is strongest in arable farming, especially cereal growing, with its relatively low operational and capital costs and its low returns per acre, where the incentive to use specialised machinery, such as combines, to maximum capacity and to spread fixed costs over larger acreages is strong, and where the occupiers of such holdings, which are already large, also tend to be among the more progressive farmers and better able to finance farm development. The proportion of the crop and grass acreage in holdings of 300 acres and over appears to have risen from 40 to 50 per cent. in the Eastern Region of the Ministry of Agriculture between 1944 and 1965, compared with from 24 to 33 per cent. in England and Wales as a whole, and the difference between the percentages in the two years exceeds 10 per cent. in most of east and south-east England, compared with under 5 per cent. in most of Wales. Scotland shows a similar, though less marked, contrast. At the same time, the number of small holdings and their share of farmland are declining. This tendency for the number of large holdings and their share of farmland to increase will probably continue, although management problems arise with large farms and very large farming empires, covering thousands of acres, have not generally survived their founders.

Finally, an analysis of regional difference in farm size should mention land in multiple occupation, i.e., common land in England and Wales and common pastures in Scotland. Common land in England and Wales is land in private ownership over which certain individuals have rights of use, especially the right to graze livestock, usually by virtue of occupying a particular holding or building. The present area of some one and a half million acres (about 5 per cent. of the agricultural land) is the remnant of a much larger acreage which was enclosed in the nineteenth century or earlier. Much lies in the uplands of south-west and northern England and Wales and is mainly rough moorland, undivided by boundary fences and providing rough grazing for sheep, cattle and ponies. The status of such land is often uncertain and it seems likely that some of the common land recorded in northern England is simply unfenced land owned in separate parcels by different individuals and grazed in common by their livestock, although the agricultural, as opposed to the legal, significance of the differences is small. At present the acreage of common land used for agriculture is estimated by officers of the Ministry of Agriculture and included in the agricultural returns. The extent of common land varies, ranging from 0 per cent. of the agricultural land in Northamptonshire to 37 per cent. in Breconshire (Figure 5). So too does the proportion that common land represents of all rough grazing, which ranges from nothing to 74 per cent. in Hampshire (mainly in the New Forest), compared with an average of 29 per cent. in England and Wales. Reasons for some of the differences are obscure and there are marked discrepancies between the acreages recorded in a Parliamentary return of 1873 and those recorded today, e.g. over 100,000 acres in Montgomeryshire in 1873 compared with under 10,000 acres now.

In Scotland no such common land survives, but in the seven crofting counties of the north-west there are nearly $1\frac{1}{4}$ million acres of common pasture on which crofters are entitled to graze stock. This land, too, comprises rough grazing in private ownership and the differences between such pasture and the common grazings south of the Border are of legal rather than practical importance.

The significance of common land and common pasture is three-fold. First, the land cannot easily be improved or change its use, even if suitable for agricultural improvement or for afforestation, for agreement by those enjoying common rights must be secured. Secondly, common land in effect enlarges the size of the holdings to which common rights attach; if the common land in Breconshire, for example, were subdivided among the occupiers of agricultural holdings, it would increase the average size of agricultural holdings in the county by 70 acres. Thirdly, since responsibility is divided and rights often uncertain, common land is liable to misuse, especially by overstocking.

THE SIZE OF BUSINESS

Acreage is only one possible criterion for measuring the size of a farm and is adopted partly because people think of farm size in this way and partly because data for the study of regional differences have been available only on this basis until recently. The size of farm can also be measured by the size of its labour force, by the capital invested in it, and by the value or volume of output. The results of such analyses often differ from those based on acreages, although there is a general correspondence among farms of the same type. The chief differences arise over comparisons of intensive and extensive farming systems. An intensive system, such as horticulture or poultry farming, may use little land, but involve heavy investment in buildings, employ much labour and have a large turnover, so that output per acre is high. In extensive systems, such as the rearing of hill sheep, relatively little capital is invested in buildings, and there are few employees and a low output per acre. Thus a horticultural holding of 70 acres, employing 6 regular workers and having a gross output of £18,000 represents, by agricultural standards, a large business, while a hill sheep farm of 3,000 acres, with a gross output of £6,000, is quite small. Until recently, there has been little comprehensive information about size of business, but the advent of digital computers in the 1950s made it possible to study business size by calculating standard labour requirements and output from census data on crop acreages and livestock numbers, using a procedure discussed in Chapter 14.

Tables 3 and 4 show how these different approaches affect the proportions of large, small and medium-sized holdings in the regions of Great Britain. In England and Wales (Table 3), where regional data are available on both the total man-days and the acreages of crops, grass and rough grazing in different size groups, the results are broadly similar, although the man-day and the acreage size groups cannot be equated and the

TABLE 3
SIZE OF HOLDINGS IN THE REGIONS OF
ENGLAND & IN WALES, 1966 AND 1967

	Eastern	South-east	East Midlands	West Midlands	South-west	Yorks and Lancs	Northern	Wales
Man-day size groups	Percentage of Man-days in Each Size Group in 1966							
–275	5	4	7	11	8	8	6	14
275–599	10	10	16	20	20	20	18	31
600–1,199	16	18	26	32	32	32	37	32
1,200–	69	67	51	37	40	40	39	22
ALL	100	100	100	100	100	100	100	100
Acreage size groups	Percentage of Acreage in Each Size Group in 1967							
–50	8	8	7	11	10	12	4	12
50–99	8	9	10	15	15	16	9	19
100–299	28	32	34	47	43	38	39	44
300–	56	51	49	27	32	34	48	28
ALL	100	100	100	100	100	100	100	100

Source: Ministry of Agriculture.

contributions of the various types of farms may be quite different; the main difference is the greater regional contrasts in the importance of large businesses. Maps prepared by the Ministry of Agriculture from a sample of holdings to show the areas where different sizes of business predominate broadly resemble Figure 5, especially the crops and grass version, in that large businesses (1,200 man-days and over) account for the largest proportions of man-days in eastern England and small businesses (275–599 man-days) in the Pennines, Wales and south-west England.

In Scotland (Table 4), where only average values are available, the differences between the approaches are more marked, reflecting greater regionalisation of farming types in

TABLE 4
AVERAGE SIZE OF HOLDING AND FARMING TYPE IN SCOTLAND, 1962

Average per full-time holding	Scotland	Highland	North-east	East-central	South-east	South-west
		Average Size of Holding by Regions				
			Man-days			
Man-days	1,152	881	861	1,494	1,544	1,257
			Total acreage			
Acres	478	1,895	191	457	501	347
			Crops and grass acreage			
Acres	140	93	115	169	231	138

	Hill Sheep	Upland	Arable rearing and feeding	Cropping	Dairy	Horticultural
	Average Size of Holding by Farming Type					
			Man-days			
Man-days	953	882	916	1,541	1,498	1,516
			Total acreage			
Acres	4,093	774	168	229	228	34
			Crops and grass acreage			
Acres	56	135	150	211	160	30

	Scotland	Highland	North-east	East-central	South-east	South-west
	Average Size of Holding on Hill Sheep Farms, by Regions					
			Man-days			
Man-days	953	936	597	1,069	1,036	912
			Total acreage			
Acres	4,093	6,086	6,201	4,318	2,157	2,150
			Crops and grass acreage			
Acres	56	41	21	70	72	65

Source: Department of Agriculture and Fisheries for Scotland.

Scotland (see Chapter 14). This is confirmed by the very similar averages for the types of farm and for the regions in which they predominate, although, as the figures for hill sheep show, there are regional differences in size within any one type.

THE SIGNIFICANCE OF FARM SIZE

The chief agricultural significance of farm size is its relationship to type of enterprise and to the viability of farms. Many smaller holdings whose occupiers are wholly dependent on them have too few acres to provide even as high an income as could be obtained from paid employment on farm work. Unfortunately, many such non-viable holdings lie on land of below-average quality, especially on the hill margins, and the enterprises which could be added to intensify production and so raise income from the same acreage (dairying, pig and poultry keeping, horticulture) are either unsuited to the environment, or already face conditions in which production satisfies or exceeds domestic demand, or are becoming increasingly concentrated in large units requiring considerable capital, as with poultry. In such circumstances, satisfactory incomes can be got only by enlarging the size of farm and, since all land is farmed, this is possible only if some farmers give up farming. Under the provisions of the 1967 Agriculture Act, the government can encourage voluntary retirement, for many farmers in such areas are elderly.

Studies of enterprises in relation to size of holding have shown that some enterprises are more characteristic of small holdings and others of large; this is particularly true in horticulture, but it also applies to farm enterprises.

Table 5 makes it clear that the smaller holdings have more than their share of intensive enterprises. This is confirmed by a comparison with their standard labour requirements; the latter shows that cereals are equally unimportant on small acreages and small businesses, but that few dairy cattle are on holdings that are small as businesses (although these often occupy small acreages). For small holders often compensate for lack of land by purchasing feeding stuffs, i.e. by using the land of other farms, or by applying more labour to the land. There are some advantages in small size for enterprises such as market gardening, which cannot easily be mechanised or those where individual attention is important. The enterprises characteristic of large holdings are those that lend themselves to mechanisation and require large acreages for the efficient use of other factors of production, especially those such as combine harvesters where the unit of supply is large. The most difficult problems arise where only extensive enterprises, such as livestock rearing, are possible.

Regional differences in holding size may have other effects. Small holdings are often more intensively farmed and produce a higher output per acre but they offer less scope for experiment and are more resistant to change; the range of enterprises is limited by the requirement of intensiveness and, because of their smaller size, their occupiers (who tend to be the less able and less technologically knowledgeable farmers) are often less willing to try new methods and enterprises. Conversely, the larger farms are often occupied by more progressive farmers, who are better qualified, are readier to accept change and have the resources to do so. Of course, much depends upon the type of farm and its environment; there is, for example, less scope for flexibility in operating a large hill sheep farm.

TABLE 5
PERCENTAGE DISTRIBUTION OF ENTERPRISES BY SIZE OF HOLDING
ENGLAND AND WALES, 1966 (Crops and grass)

Acres of crops and grass	Crops and grass	Cereals	Dairy Herd	Breeding Pigs	Laying Fowl
–20	4	1	3	20	34
20–99¾	22	11	39	25	34
100–299¾	40	38	42	25	22
300–	34	50	16	30	10
ALL	100	100	100	100	100

England and Wales, 1965 (standard man-days)

Man-days	Crops and grass	Cereals	Dairy herd	Breeding Pigs	Laying fowl
–275	11	5	6	16	10
275–599	19	14	24	23	16
600–1,199	29	28	35	25	21
1,200	42	53	35	36	53
ALL	100	100	100	100	100

Source: Ministry of Agriculture.

There is also evidence of a positive correlation between efficiency and farm size; thus, data from the Farm Management Survey in England and Wales for 1966 show a higher gross output per £100 of input on large farms (£122 for farms of 1,800 standard man-days and over) than on small (£102 for farms with between 275 and 449 standard man-days). The amount of unproductive time spent on travel around the farm tends to increase with farm size and supervision becomes more difficult on large holdings, but against these handicaps, the occupiers of such holdings are often better farmers and the larger farms are generally better laid out, with bigger fields and less fragmentation; thus, in the National Farm Survey, only 49 per cent. of holdings with between 25 and 100 acres of crops and grass had a good layout, compared with 64 per cent. of those with 700 acres and over. Other resources are also more efficiently used. Both the number of tractor hours per year and that of acres per tractor are higher on large farms; for example, it was found that on farms with 6 or 7 tractors the acreage cultivated per tractor was twice that on farms with only one tractor. A larger acreage may justify larger machines and these often embody more advanced technology and are more efficient. Furthermore, the level of mechanisation seems to be lower on small farms, since the scale of operation does not justify expenditure on sophisticated machinery. Similarly, labour can be more specialised (hence more skilled) with a large labour force and the supply of labour can more closely match the demand for it. Of course, some of these disadvantages of small size can be overcome by co-operation and by contracting, but co-operatives are little developed for farm operations, although they are important in the purchase of inputs and sale of produce. Contract work, on the other hand, seems to be increasingly important.

LAND TENURE

The way in which the land is held is also a factor in determining how it is used, though less so than formerly; for tenants have gradually been liberated from many restrictions under which they farmed a century ago. Various measures have made tenant farmers free to crop arable land as they wish, although covenants restricting the ploughing up of grassland may still apply; tenants have also become entitled to compensation for improvements they undertake.

There are two main classes of agricultural tenure in Great Britain, owner-occupation and tenancy, and three special categories, common land, crofting tenure and statutory small holding. The two major categories are not, however, exclusive, since a farmer may both own land and rent land from other landowners. Such mixed tenures accounted for 22 per cent. of the agricultural land; thus, in 1960, 9·4 per cent. of the occupiers of holdings in England and Wales mainly owned their land and rented some, and 6·2 per cent. mainly rented their land, but also owned some.

At the 1960 World Agricultural Census, 53 per cent. of the land in sole occupation, i.e. excluding common land, was farmed by owners, but the proportion varied with size of holding and from region to region. Proportions are highest on the largest and the smallest holdings; in 1950, for example, 56 per cent. of the land in holdings of under 15 acres of crops and grass was owner-occupied and 51 per cent. of land in holdings of 1,000 acres and over, compared with an average for England and Wales of 38 per cent. Many of the smaller holdings have owners with other sources of income; the large holdings also include the home farms of large estates. Owner-occupation is commonest in Scotland, Wales and to a lesser extent in south-west and south-east England, while tenants survive most strongly in the Midlands and northern England (Figure 5). These conclusions must be interpreted with caution as owner-occupation may refer to a large home farm on an estate which otherwise comprises tenant farms; this explains the high proportions of owner-occupation in the Scottish Highlands, where large acreages are in hand. Estates have been more stable in Scotland than in England and Wales and there are fewer owner-occupiers in the sense of a man owning only one farm which he manages. Tenant-farming is, indeed more strongly established in Scotland, where there are rights of succession by close relatives on the death or retirement of a tenant. In Wales, on the other hand, holdings owned by occupiers are quite small and result from the break-up of large estates.

This situation contrasts markedly with the mid-nineteenth century, when, apart from home farms, most farmland was occupied by tenants. Death duties, taxation and other reasons have led to the break-up of large estates and often to the purchase of farms by sitting tenants who would have had to be compensated for improvements if dispossessed. Since 1945 an added reason has been the difficulty of acquiring a tenanted farm because of the security of tenure enjoyed by sitting tenants, so that would-be farmers with the necessary resources have had to buy to obtain a farm. The unprofitableness of renting land has also encouraged landlords to take land in hand at the end of a tenancy and farm it themselves. Greater interest in part-time and hobby farming is a further factor; those with employment outside agriculture have been willing to buy a farm, regarding it rather as a home than as a business. Some land purchase also contains a speculative element, especially near large cities. The proportion of occupiers is unlikely to rise much higher, for much of the remaining agricultural land is held either by institutional landowners, who have purchased it as an investment, or by official and semi-official bodies.

Crofting tenure, a special form of tenant-occupation which provides considerable security of tenure and rent control, has been described as giving all the advantages of land ownership without any of its disadvantages and applies only in the seven crofting counties of Argyll, Caithness, Inverness, Orkney, Ross and Cromarty, Sutherland and Zetland. Crofting tenure was established under the Crofters' Act of 1886 to protect the small tenantry surviving there and to perpetuate certain traditional privileges. A croft is a holding not exceeding 75 acres (exclusive of any common pasture) or with an annual rent of £50 or less, and a Crofters' Commission controls letting and promotes the re-organisation of crofts. There are now about 20,000 crofts, occupying over 150,000 acres of crops and grass, and crofting is most prominent on the coastal fringes of north-west Scotland and on the islands, especially the Outer Hebrides, where over half the improved land is in crofting tenure and over half the rough grazings are common pastures; away from the north and west coasts there are few true crofts. Few of these holdings can provide their occupiers with a living and many are tenanted by elderly crofters or even by absentees, some of whom have even emigrated; such crofts are often neglected.

Another special form of tenant-occupation is the statutory small-holding. In England and Wales these have been established under a number of acts since 1892 which have given powers to county councils to establish small-holdings. There are about 15,200 such holdings covering some 430,000 acres, the proportion varying greatly from county to county. In Scotland, the Department of Agriculture and Fisheries owns some 4,000 holdings, which also cover about 430,000 acres, and hence are very much larger; 90 per cent. of them lie in the crofting counties. Statutory small-holdings were initially established to enable landless labourers to become full- or part-time farmers, to resettle ex-servicemen after the First World War and to help the unemployed in the depression of the 1930s. Only about half of them are now capable of providing full-time employment; in 1968 a Committee of Inquiry in England and Wales recommended that their number should be greatly reduced to provide some 5,000 viable small-holdings as a 'gateway' for the best new entrants into farming.

Less is known about the pattern of land ownership and about its significance for agricultural purposes. According to a sample survey by D. R. Denman, about 24 per cent. of land is owned by public bodies and a further 30 per cent. by institutional land owners; private individuals own the remainder and range from the Duke of Buccleuch, with 222,000 acres, and the Countess of Seafield with 213,000 acres, to the man owning a few acres of land. There is however, little accessible information about the distribution of such estates, apart from the Highlands, of which R. Millman has published a map.

There is not much firm evidence about the *agricultural* significance of land tenure. Traditionally, the landlord-tenant system was a partnership, with the landlord providing the fixed equipment and the tenant the working capital, but the situation has been greatly changed by the growth of owner-occupation and by legislation to regulate the relationship between landlord and tenant. Owner-occupiers enjoy greater freedom in farming, even to the extent of neglecting the land if they wish, and possibly farming is more varied in areas where owner-occupation predominates. But, unless such owners are wealthy, they have tied up much of their capital in obtaining a farm and hence have less for farming operations; indeed, owner-occupiers sometimes sell their land to become tenants again. In law tenants have considerable freedom, but in practice an estate can still influence the way in which tenants farm, both directly through the provision of building and other improvements, and indirectly through the expression of approval or disapproval. From the outbreak of the Second World War until 1958, tenants enjoyed great security of tenure, although there were rarely-used powers for the dispossession of those, whether tenants or owner-occupiers, who did not farm their land properly. Since 1958, the government cannot dispossess occupiers, but the grounds on which notice to quit can be given by landowners have been enlarged and the freedom of tenants has been correspondingly diminished, although both tenant and owner have the right of appeal to the Lands Tribunal or to the Scottish Land Court. The attitudes of landowners towards their land may also affect the pattern of agriculture, especially where they are influenced by non-agricultural consideration; for example, the Ministry of Defence controls the agricultural use of large tracts used as military training grounds, and on large private estates conflicts may arise between farming and management of the land for game birds.

FARM LAYOUT

Shape and layout also affect the way in which agricultural land is used, although their effects are mainly local, except in so far as they have regional characteristics. Four elements are relevant: the shape of farms (both in a general sense and in relation to the terrain), the degree of fragmentation, the nature and location of farm buildings, and the size and shape of fields.

No systematic information is available about the shape of farms, although some could be obtained by sampling the farm boundary maps. The one certain fact is that they are irregular, unlike the grid-iron pattern in North America; this is so even in central Scotland and in the English lowlands from the North Riding to Dorset where new farms were laid out in the eighteenth and early nineteenth centuries, although boundaries here are more frequently rectilinear. Ideally, a farm should be compact, with all parts equally accessible from the farmstead. There are also advantages, at least on large farms, in having a range of physical conditions within the farm. Such an arrangement might provide early and late grazing (where a variety of aspects was included), help to spread the harvest, and provide some insurance against the extremes of drought and overabundant moisture; on the other hand, the trend towards simplification of farming systems suggests that the range of conditions should not be great. The most distinctive shape of farm is associated with areas where land of different types occurs in parallel bands, as on the scarplands, or along the coastal marshes, or where valleys penetrate the uplands; in such circumstances it is common, although not universal, for farms to be long and narrow and to include a variety of terrain. Farms in areas of late enclosure are also sometimes elongated, with the farmstead in the village.

Fragmentation of farms is rarely a serious problem in Great Britain. It is most characteristic in three situations; where farm enlargement is proceeding rapidly, where there are many small holdings, and where parliamentary enclosure occurred late. Apparent fragmentation, i.e. a holding in several discrete parts, often indicates farm expansion. Land becomes available only infrequently, generally on a farmer's death or retirement, and land adjacent to an existing holding may not be available, although these fragments (generally substantial) may ultimately be linked by the acquisition of the intervening land; in a group of parishes in Nottinghamshire half the holdings which increased in size between 1942 and 1964 became more fragmented as a result. Fragmentation in small holdings poses more serious difficulties; it is a major factor in several problem areas in the lowlands and is difficult to alleviate without legislation, as the failure to reorganise farm boundaries in the Yetminster area shows. Fragmentation arising from late enclosure is especially characteristic of Midland counties. The only source of information about the regional importance of farm layout is the National Farm Survey, which was confined to England and Wales, and showed that, while fragmented holdings accounted for 25 per cent. of all holdings in the twenty-four counties surveyed, the proportion of severed holdings ranged from under 5 per cent in the south-west peninsula to over 50 per cent. in Cambridgeshire and Nottinghamshire, where much enclosure occurred late and fragmentation is often associated with village farms.

The location of farm buildings is related partly to the history of settlement and enclosure and partly to recent trends in farm size. All the farm buildings are normally located at the farmstead, although sometimes barns and other buildings are scattered throughout the fields, as in the Pennine dales. The farmstead may be located centrally on the holding or at its periphery, or even away from the holding, and may be isolated from other settlements or included in a larger settlement. Isolated farmsteads are most common in Scotland, Wales and much of western and northern England, areas of mainly dispersed settlement. Where the village is the characteristic form of settlement, however, there are often many farmsteads in the village itself. In areas of late parliamentary enclosure, steadings, which had been located in the village when open-field farming was practised, were often rebuilt away from the village after enclosure, except where villages were small and moving was unnecessary. Where villages are large, many farm buildings, especially on small holdings, are still located in villages, separated from the farm territory or connected with it only by a narrow strip of land. Apart from the farm boundary maps, there is no general source of data, although a survey of the Chilterns and the clay vales has shown that, while nearly all Chiltern farmsteads are isolated, about half those in the clay vales lie in villages, although the proportion varies from village to village. Cottages, too, may be scattered throughout the holding, located in a village or sited near the farmstead, as is customary in eastern Scotland.

Farm buildings, too, vary greatly in character, from the black house of the Hebrides to the massive stone courts of the Lothians and the modern factory farms that are increasingly common in eastern England, although the typical farmstead, especially in eastern counties, is a heterogeneous collection of buildings of different ages. Farm buildings represent an increasingly important investment in land, and improvements are facilitated by government grants; indeed, over 80 per cent. of the investment under the Farm Improvement Scheme between 1957 and 1962 went into buildings. Most farmsteads were constructed at least one hundred years ago, generally for quite different types of farming, and although they have been adapted piecemeal, both by modifying existing buildings and by constructing new, they are often ill-suited to present needs and may well impede the adoption of alternative systems of farming. In 1949, for example, with the introduction of Dairy Regulations, some milk producers in Caernarvonshire had to abandon production because their buildings were unsuitable.

The last element in farm layout is the component fields. Although there is abundant material for assessing regional differences in field size on the large and medium-scale maps of the Ordnance Survey, it has never been systematically analysed. Considerable regional differences in field size are, of course, observable throughout the country, although any survey immediately poses the question 'What is a field?' If a field is an area of agricultural land surrounded by a fence, wall, hedge, ditch, or any combination of these, fields on enclosed land in western counties are characteristically small, generally under 10 acres and often under 5, and those in arable districts are large, exceeding 100 acres in places, especially on the chalk plateau of southern England. Thus, on a Devon farm where the merits of field enlargement were being investigated, the largest field was 8·7 acres and 20 of the 38 fields were under 3 acres, while in North Hertfordshire, on the arable land below the chalk escarpment, 33 per cent. of the fields, covering 74 per cent. of the land, were larger than 30 acres. This impression receives some statistical confirmation from measurements by G. M. L. Locke of the lengths of field boundaries in selected counties, which suggest, for example, that on average fields in Essex are four times as large as those in Devon. The national fertiliser survey also provides data, although the interpretation of 'field' appears to be very flexible; these calculations are only a by-product of the main purpose of the survey and must be viewed with caution. The data from sample counties given in Table 6 also confirms general impressions about the distribution of fields of various sizes. The differences are related partly to the evolution of the agrarian landscape, for early enclosure often produced small fields, and partly to type of farming. Both in Holland, with widespread horticulture, and in Cheshire, with dairy farms predominant, the land is mostly in fields of under 20 acres, whereas in Berkshire, with much large-scale cereal-growing, over half is in fields of 30 acres and above and over a quarter in fields of 50 acres and above. This regional distribution is also partly related to farm size, for small fields tend to be associated with small farms and large fields with large farms.

Although fields in eastern and southern counties tend to be much larger than those in the north and west, there are many local variations. In sample areas of north Hertfordshire, for example, fields of under 10 acres comprised only 9 per cent. of agricultural land, but 66 per cent. in a sample area of south Hertfordshire, where no field was larger than 25 acres. There is also a wide range of types, for field shapes are generally regular in areas of late enclosure and irregular elsewhere; but no systematic studies have been made of regional differences in field shape.

TABLE 6

PERCENTAGE OF LAND OCCUPIED BY FIELDS OF DIFFERENT SIZES
(crops & grass)

Area	Size Groups (acres)							
	Under 10	10–14	15–19	20–29	30–39	40–49	50 and over	All
Berkshire	4.4	8.7	9.8	20.6	17.5	12.0	27.1	100.0
Lincolnshire	10.2	12.0	13.1	21.1	12.7	6.0	25.0	100.0
N.E. Northants	12.5	19.4	15.5	22.9	9.2	7.0	13.3	100.0
Holland	37.7	17.7	12.8	18.8	5.8	3.1	4.0	100.0
E. Shropshire	24.8	24.3	16.8	19.5	7.1	3.6	3.8	100.0
C. Cheshire	40.4	29.7	14.7	8.2	4.2	1.4	1.3	100.0
N. Lancashire	40.4	18.0	9.4	19.2	5.1	4.5	3.3	100.0
N. Devon	56.6	22.5	11.4	4.5	3.0	0.0	1.9	100.0
N.W. Denbigh	55.2	20.8	9.0	7.3	2.8	2.0	2.8	100.0
W. Carmarthen	66.4	15.5	9.0	5.1	1.9	1.4	0.7	100.0
ALL SAMPLES	31.8	18.9	12.3	15.6	7.4	5.0	9.0	100.0

Source: National Fertiliser Survey.

As with farm size, there are strong forces making for the enlargement of fields; in a sample area in Huntingdonshire, the length of field boundaries, which is one measure of field size, was reduced by 70 per cent. between 1944 and 1964. The pressures are strongest where mechanised farming is most highly developed and there are few grazing stock, i.e. in eastern arable districts, and most resisted where field boundaries are massive, e.g. the earth banks of Devon or the shaws of the Weald, and the costs of removal considerable—£300 a mile in one sample study (although tax relief and government grant would reduce the net cost to nearly £110). Enlargement brings benefits in easier and more efficient operation and in the release of land for crop production; but it is deplored by conservationists, who consider that a proportion of wild land (in hedges, copses and the like) is necessary for a healthy countryside and allege that removal of trees and hedges is partly responsible for an apparent increase in wind erosion in eastern counties.

There is little doubt that there are agricultural advantages in field enlargement. Pre-war labour costs of wheat growing in East Anglia were 11 per cent. lower in fields of over 15 acres than in fields of under 10 acres, and in one Devon example the time required for combining was reduced by a third after field enlargement. Increased size of machinery and speed of operation make larger fields desirable on many arable farms, although whether there are advantages in very large fields is unproven. On the other hand, hedgerow removal reduces flexibility and makes the keeping of grazing livestock more difficult by destroying shelter and stock-proof fences.

The National Farm Survey of 1941–3 represents the only attempt to measure farm layout systematically, and its significance for agricultural activity thus remains largely unexplored, although the rapidity with which both fields and farms are being enlarged suggests that farmers see very real advantages in increasing the size of both. Undoubtedly farming efficiency would be improved with a more rational layout of farms and fields, and R. N. Dixey and A. H. Maunder estimated in north Oxfordshire that an improved layout would lead to an increase in farm profit from £2·4 to £5·3 per acre. To some extent the trend towards larger farms will itself lead in time to better farm layout and the Agricultural Land Service, in its land classification map of south-east England, used the percentage of holdings over 300 acres as an index of good farming structure which would justify the upgrading of land into higher categories. The desirability of improving farm layout is accepted in government policy and grants are available under the Farm Improvement Scheme towards the costs of field enlargement and other improvements, while the 1967 Agricultural Act authorises payment of grants towards the cost of farm enlargement.

The fixed equipment of farms is of considerable importance. A. L. Langdon has shown that in sample farms in the West Midlands the replacement cost of the buildings was little less than the whole value of the farm, suggesting that bare land, so far as this is conceivable in a British context, was of little value in itself. Investment in fixed equipment is increasing, but, important as it is, it cannot easily be taken into account in assessing the importance of agricultural production owing to lack of suitable data.

LAND VALUES

If land quality subsumes the physical qualities of land, land values provide a rough index of the quality of land in the wider sense (Figure 2). As with so many aspects of the physical and human environment of farming, satisfactory data are lacking. Average rent is probably the most useful index, although, with half the agricultural land in owner-occupation, such rents must often be notional; county rent levels show a wide range of values, from over £5 per acre in the Fenlands and in Cheshire to under 1s. in Sutherland. Location, land quality, type of farming, size of farm, and many other factors contribute to these variations, and rents are generally highest in areas of good quality land and of intensive systems of farming, especially dairying and horticulture. In any one area they are highest on small holdings (where the farmhouse and steading represent a high proportion of total value) and lowest on large farms, where the fixed equipment charges are spread over a large acreage; thus, according to the National Farm Survey in 1941–3, the rent of holdings between 5 and 25 acres was 52s. and that of holdings of over 700 acres 19s. Although rents have risen sharply since 1958, they are still low in relation to land values: while average rents have risen threefold since 1939, the sale value of agricultural land with vacant possession has risen eightfold and for land without vacant possession sevenfold.

Economists have emphasised the declining importance of agricultural land and this is true in a relative sense; for the proportion of expenditure on United Kingdom farms represented by rent and rates fell by more than a third between 1937 and 1967. The proportion varies both relatively and absolutely with size and type of holding; thus, while the average for the United Kingdom was 14 per cent. in 1967, the proportions on different types of holding ranged from 7 per cent. (average value £11·8 per acre) on horticultural holdings to 20 per cent. (average value £5·1 per acre) on cropping, mainly cereals, farms. Rent, whether actual or notional, is thus a substantial fixed cost which must be met, and farmers are giving careful consideration to the relationship between rent and the net income they can derive from particular enterprises; on cereal farms, for example, a rent of £5·1 per acre compares with a net income of £8·8. Nevertheless, labour and machinery are both more important items, accounting for 31 per cent. and 26 per cent. of expenditure respectively; to these other major inputs we shall now turn.

FURTHER READING

J. T. Coppock, 'Farms and fields in the Chilterns', *Erdkunde* 14 (1960)

E. T. Davies and W. J. Dunsford, *Some Physical and Economic Considerations of Field Enlargement*, Rept. No. 133, Dept. of Agric. Econ., Univ. of Exeter (1962)

D. B. Grigg, 'Small and large farms in England and Wales', *Geography* 48 (1963)

R. N. Dixey and A. H. Maunder, 'Planning again; a study of farm size and layout,' *Farm Econ.* 9 (1958–61)

R. Millman, 'The marches of the Highland estates', *Scot. Geogr. Mag.* 85 (1969)

University of Newcastle, Agric. Adjustment Unit, *Farm Size Adjustment*, Workshop Report, Bull. No. 6 (1968)

University of Nottingham, Dept. of Agricultural Economics, *Farm Amalgamation 1950–64*, F.R. 159 (1965)

CHAPTER 5

Men and Machines

Agriculture in many developed countries combines a high input of manual labour with heavy investment in machinery. Although over the past century men have become fewer and machines more numerous in British agriculture, the latter particularly in the past three decades, there are quite modest limits to the extent to which machinery can be substituted for labour in the predominantly livestock economy of British agriculture as at present organised. Many farm tasks cannot easily be mechanised, although the agricultural engineer is increasing the range, which will be further enlarged as the geneticist works more closely with him to produce crops better adapted to mechanical handling; the structure of farms (Chapter 4), especially the small size of many holdings, is a further obstacle to mechanisation. These difficulties in mechanising agricultural production are partly responsible for the persistent disparity between industrial and agricultural wages. Although neither labour nor machinery has been the subject of geographical enquiry, certain characteristics of these two major agricultural inputs will be examined in this chapter, particularly those features likely to contribute to our understanding of the varied pattern of British agriculture.

LABOUR

Like many features of British agriculture, the nature of farm labour is difficult to define; even such questions as 'What is a farmer?' and 'What is a farm worker?' present considerable problems in practice, for those who work on the land form a continuum of varying degrees of dependence on and involvement in agriculture. Is a business man who runs a farm as a hobby, leaving the detailed supervision to a farm manager, a farmer? Is a farmer's wife, who keeps a flock of chickens and helps at harvest-time, a farm worker? Even where such questions can be satisfactorily answered, data are often lacking, particularly at regional or county level.

In a national perspective only a very small proportion of the country's labour force is employed in agriculture (although the actual number depends on where the line between agricultural and non-agricultural employment is drawn). The census returns for 1966 (based on a 10 per cent. sample) recorded a total of 683,770 employed in agriculture, or rather under 4 per cent. of the total labour force. Most of the remainder are, of course, employed in manufacturing and in the service industries and are mainly town dwellers, but even in the rural areas the proportion employed in agriculture rarely exceeds 50 per cent. Such people, together with those employed in other rural industries such as quarrying and forestry, constitute the primary population; the remainder of the rural population comprises either those ministering to the needs of farming communities, the service or secondary population, and what L. D. Stamp termed the 'adventitious' population, that is, those who live in the country from choice either because they have retired or are of independent means, or because they work in towns but prefer to live in the country. Adjustment of administrative boundaries lags behind urban growth and some rural districts include areas which are strictly urban; yet, even when allowance has been made for such anomalies, the adventitious population depresses the proportion of the working population employed in agriculture around the large towns and cities and in some favoured areas of attractive scenery, such as the Cotswolds. Only in flat lowlands away from towns and in the uplands does the proportion of agricultural workers approach S. W. E. Vince's calculated norm of two-thirds in primary employment and one-third in secondary. These are the truly rural areas. Elsewhere, agriculture functions in an urbanised environment the so-called 'rurban' areas, where recreational pressures are generally greatest and trespass and damage by the urban population often a severe handicap to farming (although these problems may also affect areas of high amenity, such as the Lake District). But the *density* of the agricultural population is not necessarily highest in the rural areas proper, for these include both uplands, with fewer than 10 people employed in agriculture per square mile, and areas like the Fenland, where the agricultural labour force exceeds 100 per square mile; indeed, the highest densities are on the outskirts of the great cities where intensive horticulture is practised, despite these handicaps, as around London and in south Lancashire.

This labour force varies greatly in composition, depending on the degree of specialisation of function, on contractual status as self-employed, employer, employee or unpaid relative, and on the proportion of the working year devoted to agriculture. The resulting divisions are not clear-cut and the same individual may legitimately be included in different categories; for example, a man may be self-employed and a part-time farmer may be both manager and labourer.

Farm work covers many functions, entrepreneurial, managerial and manual (the latter including operations requiring a wide variety of skills). The occupier of a holding must make decisions about the kind of farming to be undertaken, the combination of enterprises in the farming system, the kind and quantity of different inputs, and the markets for the produce; decision-making of this kind is inherent in operating a holding, even if decisions are not consciously or rationally made. Too little is known about such decision-making and about the motivation and aims of occupiers, but undoubtedly these are only partly based on economic considerations, about which the farmer may, in any case, be ill-informed. Functions may be combined to varying degrees in one person or in several, as in a partnership, or separated, as on holdings with a full-time manager. Most farmers are both entrepreneurs and managers, implementing the policies they have decided and organising their resources to achieve these ends. On large farms or where occupiers have their main employment outside agriculture, management may be by a full-time manager, and there were estimated to be 12,000 such holdings in England and Wales in 1960, (although 47 per cent. of these were under 125 acres); it is interesting that, in the Farm Management Survey sample, the highest figure for salaried management relates to farms on the Cotswolds and the chalk downs. Part of the day-to-day management may also be delegated to foremen, grieves or bailiffs, who accounted for 6 per cent. of all employees in 1966;

MEN AND MACHINES

in England and Wales they were nearly all on holdings with two or more regular male employees and only 12 per cent. were on those with fewer.

Most manual work in agriculture is skilled, whether it be managing stock of various kinds, operating a wide range of tillage and harvesting machinery or undertaking maintenance such as repairs, hedging and ditching. This work can likewise be divided amongst specialists or combined to varying degrees in one or more persons; on many smaller holdings a high degree of specialisation is not possible unless the holding itself is specialised. Most occupiers of small holdings combine all these functions, but the proportion of time devoted to manual labour falls with increasing acreage.

In Great Britain, 35 per cent. of regular full-time male workers were classified as general workers in 1966. In Scotland specialists predominate, but in England and Wales, on the other hand, there were 95,300 general farm workers in 1965, compared with 15,100 dairy cowmen, 19,900 stockmen and 33,700 tractormen. Understandably, the proportion of general workers decreases as holdings get larger; on holdings with under two regular whole-time male employees there are more general workers than specialists, while the reverse is true on large holdings. Thus, in 1965 tractormen formed 23 per cent. of the adult male labour force, on holdings with 10 or more regular male employees compared with 9 per cent. respectively on holdings with 4 or fewer employees; for general workers the proportions were 33 per cent. on holdings with 10 or more employees, and 58 per cent. on holdings with 4 or fewer. The proportion of general workers has declined, presumably because of greater specialisation and mechanisation in farming and of the increasing size of farms; among specialists, the increase in mechanisation has produced a rise in the number of tractormen and a decline in numbers of stockmen.

The agricultural labour force also varies in age. Ministry of Labour statistics for employed labour in England and Wales show that the percentage in the older groups (55+) is unusually high and has been rising; moreover, the proportion is higher in the remoter parts of the country. In Scotland, too, the proportion of older workers is highest in the Highlands, where 7·1 per cent. of regular workers are over 65, compared with

G

3·7 in Scotland as a whole. Age distribution varies with type and size of farm and with kind of labour; family labour, for example, is younger on average than non-family labour because many sons become farmers, while young workers form almost twice as high a proportion of full-time employees on dairy farms in Scotland as on hill sheep farms. The population of farmers, too, is above average age in upland areas and, in the Mid-Wales Report, two-thirds of those surveyed were over 50 years of age.

FARMERS

The other principal basis of distinction within the agricultural labour force concerns employment status, particularly whether those who work on the land are self-employed, employers or employees, whether they devote all or part of their time to agriculture, and whether any part-time employment is regular or casual. The most obvious distinction might appear to be between farmers and farm workers, but it is not easy to say who is a farmer, chiefly because of differences in their degree of dependence on the land. The population census provides one source of information; there were estimated to be 260,000 farmers in 1966 (excluding farm managers), and densities vary with farm size, being high where farms are small, as in the Fenland, and low where they are large, as in Northumberland. Such averages may mislead, as the contrasts within the crofting counties illustrate; for small crofts occupy mainly coastal sites, while the interior is occupied by vast sheep farms.

An alternative approach uses the agricultural returns, which identify not farmers, but occupiers of agricultural holdings, many of which are not normally regarded as farms; indeed, many of the smallest are only plots of agricultural or semi-agricultural land. This difficulty can in part be overcome by estimating, from standard labour requirements, the number of holdings providing full-time employment for one or more men; the occupiers of most of these will be returned as farmers in the population census. In 1965 there were about 181,000 such full-time holdings, or 50 per cent. of the total, and their distribution corresponds broadly with that of farmers as recorded in the census. This method is least accurate at the lower end of the scale, where occupiers of nominally full-time holdings may in fact be part-time, while some who occupy holdings with labour requirements appropriate to the part-time category are wholly dependent on their holdings; a survey in 1965–7 in Scotland found that 8 per cent. of the part-time holdings appeared to be full-time, while 3 per cent. of the full-time holdings were, in fact, part-time.

The occupiers of such holdings vary greatly in their dependence on their holdings and in social and economic status. In a survey in England and Wales in 1960, some 11 per cent of the occupiers of genuine part-time holdings had no other source of income, 42 per cent. had other full-time employment and 14 per cent. had part-time employment. In a similar survey in Scotland, 14 per cent. of occupiers were wholly dependent on their holdings, 36 per cent. had full-time occupations and 10 per cent. had part-time employment, the remainder having other sources of income. In England and Wales 18 per cent. of those with full-time jobs and 31 per cent. of those with part-time jobs were in other agricultural employment, often as agricultural labourers, and thus combined more than one role on different holdings.

Few of those not employed in agriculture are engaged in other primary occupations, but considerable regional differences in the status of occupiers must be borne in mind in interpreting Figure 6, which shows the proportion of part-time holdings in each county. They are commonest in certain rural areas and also near the large cities, but the reasons for these concentrations are very different. In north-west Scotland, most occupiers of part-time holdings are crofters, and in other upland areas part-time holdings are often a legacy of mineral working in the past, as in Caernarvonshire, where 71 per cent of holdings were part-time in 1961. Around the cities, part-time holdings are often occupied by white-collar workers; in sample studies in Kent and Buckinghamshire 58 per cent. and 62 per cent. respectively were business or professional men.

We should note in passing that, even where holdings are genuinely full-time, their occupiers may not be full-time farmers. Especially in the Home Counties and in areas of attractive scenery, such as the chalk downland and the Cotswolds, they may be business or professional men, for whom farming provides both a hobby and a pleasant rural residence. The home farms of large estates may similarly be occupied by those who combine farming, often on a large scale, with estate management or with non-rural occupations. Until 1960, profits from other businesses could be offset for taxation purposes against farming losses and there were undoubtedly businessmen who intentionally farmed at a loss. This is no longer permissible, but hobby farming still retains many attractions.

For distinguishing between farmers who are self-employed and those who employ labour the population and agricultural censuses are again not wholly compatible. According to the 1961 population census, 47 per cent. of farmers were not employing labour, and the remainder were either employers of labour or managers. Most occupiers of full-time holdings are employers, but 41 per cent. of such holdings in England and Wales and 30 per cent. of those in Scotland had no regular full-time male labour (the most important category of employees).

Lastly, occupiers of holdings need not be individuals, although most are; B. E. Cracknell and J. Ashton, using the agricultural census list of names, found that 88 per cent. of holdings in England and Wales were occupied by individuals and 9 per cent. by partnerships, (although these are often between father and son, or between brothers or other relatives). Corporate bodies, such as institutions, joint stock companies and the like, formed the remainder and were most prominent in south-east England. Estimates of the proportion of land occupied by different categories of occupiers provide a somewhat different picture; 19 per cent. of the farms occupied by companies in England and Wales were of 500 acres or over, although holdings of this size accounted for only 2 per cent. of all holdings. Occupiers may also be classified as male or female, although this has no known agricultural significance; according to the population census, 92 per cent. of farmers in 1966 were male.

Differences in the status of occupiers may influence the way in which land is farmed and the efficiency of farming; and even where the land is well managed, differences in attitudes to land, to profit and to different enterprises may result. Many of the part-time holdings are poorly farmed, especially if natural conditions are difficult and the occupiers elderly, as in the crofting counties and in upland Wales; such holdings are of little agricultural importance and occupiers of 20 per cent. of those in Scotland sold no agricultural produce. In contrast, well-educated hobby farmers with adequate capital are often innovators in farming. The influence of status is revealed in Miss Gasson's interesting study of farming in south-east England: she showed that part-time farmers in an area southeast of London derived more of their output from cereals, pigs and poultry than did full-time farmers and tended to have simpler farming systems.

Even among full-time farmers differences in attitude are important. J. B. Butler, in his study of farmers in the Vale of York, found that the differences in the types of farming in a fairly homogeneous physical environment could largely be explained in this way. Some of these reflect differences of temperament and personality which have a random distribution throughout the population at large; but others, such as age, training and background, may have a regional component, as with the high proportion of elderly farmers in upland areas of Wales. Miss Gasson, also, noted contrasts in education and social background between small farmers in the Fens and those on the outskirts of London in Hertfordshire which appeared to affect significantly their attitudes to farming. Migrant farmers represent a particularly interesting and important group. The Scottish dairy farmers who moved to Essex and Hertfordshire in the late nineteenth century, the Dutch horticulturalists who settled around Hull in the 1930s and the Lincolnshire arable farmers who acquired farms in Romney Marsh in the 1940s and 1950s all brought new ideas and new enterprises which substantially modified the agricultural patterns of these areas.

FARM WORKERS

The last element in the labour force consists of employees, although complications arise from the employment of part-time, casual and unpaid family labour; for example, some of those returned as employees are, in fact, junior partners. The population census gives a total of 389,050 agricultural workers in 1961, compared with 487,969 in the agricultural census, where farm workers include all family labour other than farmers' wives and are divided on the basis of the proportion of time they devote to agriculture. Three categories are recognised, regular full-time, regular part-time and seasonal or casual (although the distinction between casual and part-time work is not very clear); these categories are subdivided by age and sex. Full-time workers are by far the most important,

TABLE 7
FULL-TIME MALE LABOUR IN EACH REGION IN ENGLAND AND IN WALES, 1967

	Eastern	South-east	East Midlands	West Midlands	South-west	Yorks and Lancs	Northern	Wales
	Percentage of holdings with labour							
	47	47	44	41	42	40	47	28
	Percentage of holdings with regular full-time male labour							
	34	35	31	28	28	27	34	16
Size groups regular F.T. male workers	Percentage of holdings with regular full-time male labour by size groups							
1	35	41	49	52	57	56	58	72
2	21	23	24	24	23	24	23	20
3–4	21	20	15	16	13	14	13	6
5–9	16	12	8	6	5	6	5	2
10–	7	5	3	2	1	1	1	—
ALL	100	100	100	100	100	100	100	100
Size groups regular F.T. male workers	Percentage of workers on holdings with regular full-time labour (by size groups)							
1	10	13	19	25	29	28	31	48
2	11	15	19	23	23	24	25	27
3–4	19	22	20	27	22	22	23	14
5–9	27	24	19	16	16	18	16	9
10–	33	26	23	9	10	9	6	3
ALL	100	100	100	100	100	100	100	100

Source: Ministry of Agriculture.

both in numbers and, *a fortiori*, in labour input; in 1966 there were 364,528 regular full-time workers, of whom 92 per cent. were male. They are most numerous in south-east England and the Fenland, the principal areas of horticulture, and densities were generally highest in those eastern counties where tillage predominates and intensive crops are grown; the lowest densities are in the uplands and in counties where livestock rearing is the major enterprise (Figure 6).

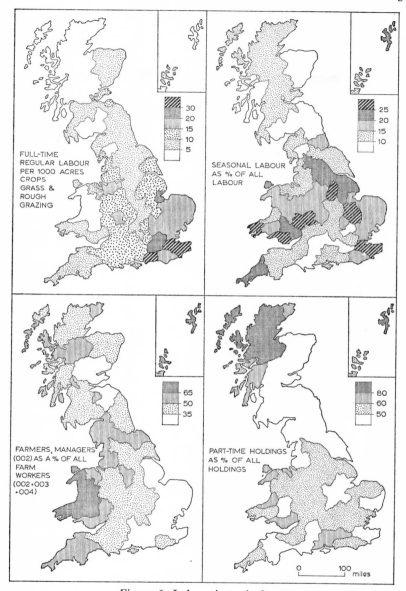

Figure 6. Labour in agriculture
Sources: 1961 population census and 1965 agricultural censuses. The numbers 002, 003 and 004 refer to census occupational categories

Tables 7 and 8 show the proportion of holdings with full-time labour by regions; they are not strictly comparable since the figures for England and Wales relate to male workers only.

The proportion of holdings with full-time regular male workers is similar to that of holdings with any employed labour, and is lowest in west and north England and in Wales, where family farms are more common (cf. Figure 6). In all regions, the proportion of the holdings with full-time regular male

TABLE 8
FULL-TIME LABOUR IN EACH REGION IN SCOTLAND, 1967

	Highlands	North-east	East-central	South-east	South-west
	Percentage of holdings with labour				
	23	37	56	58	54
	Percentage of holdings with regular full-time labour				
	9	31	52	53	46
Size groups regular F.T. workers	Percentage of holdings with regular full-time labour (by size groups)				
1	48	51	33	29	44
2	26	25	24	21	28
3–4	17	15	25	28	20
5–9	8	7	16	18	7
10–	1	2	2	4	1
ALL	100	100	100	100	100
Size groups regular F.T. workers	Percentage of workers on holdings with regular full-time workers (by size groups)				
1	22	23	11	9	20
2	23	23	16	12	25
3–4	26	23	28	28	29
5–9	21	20	34	33	18
10–	7	12	11	18	9
ALL	100	100	100	100	100

Source: Department of Agriculture.

workers in each size group decreases as the size of the labour force increases, but the proportion of the number of such workers does not necessarily do so; however, the share of the larger size groups decreases with the fall in the proportion of holdings with any full-time labour.

Similar results are revealed by the Scottish census, although the proportion of holdings with any labour is higher in east and south Scotland than in any English Region, and the range of differences is greater, a reflection in part of the greater environmental contrasts and also of the greater homogeneity of the Scottish Regions.

Part-time and seasonal labour (the two were not separately distinguished before 1955) accounted for 12 per cent. and 17 per cent. respectively, although the June figures understate the importance of the seasonal labour force, which is largest when crops are being harvested; the number recorded in 1965 at the September census for England and Wales was 23 per cent. higher than in June. Women figure more prominently in the part-time labour force (45 per cent.) and casual labour force (46 per cent.) than among full-time employees, and the regular employment of large numbers of women part-time is characteristic of some horticultural areas. Temporary workers recruited from the towns, notably for the hop and potato harvests (the latter used to employ half-a-million people), have declined sharply with the mechanisation of harvesting.

TABLE 9
CASUAL LABOUR IN ENGLAND AND WALES, 1967

Size of business Standard man-days	All types*	Specialist Dairy	Livestock Sheep & cattle	Cropping Mainly cereals	Horticulture
Casual labour as a percentage of all paid labour					
275–449	27	33	22	11	—
450–599	17	19	6	14	—
600–1,199	10	9	5	7	32
1,200–1,799	8	4	6	6	24
1,800	12	6	7	6	22
ALL SIZES	12	9	7	7	23

* excluding horticulture

Source: Farm Management Survey.

The importance of seasonal labour is related both to size of business and to type of farming. Table 9, derived from the sample farms in the Farm Management Survey, shows the proportion of the labour bill represented by casual labour on holdings of different sizes and different farming types. It is most important in small businesses and on horticultural and general cropping farms, and least on those with extensive enterprises such as cereal-growing and livestock rearing. This helps to explain the distribution of seasonal labour, which is especially important in areas of intensive cropping (Figure 6).

Little is known about the contribution of family labour, since no specific information on it is included in the agricultural censuses for England and Wales, although some data are collected in the December census in Scotland; no data are available about the separate contribution of farmers' wives. A special enquiry in connection with an annual survey on

TABLE 10
FAMILY LABOUR IN ENGLAND AND WALES, 1967
(Labour of farmer and wife as a percentage of total labour costs)

Size of business Standard man-days	All types*	Specialist Dairy	Livestock Mainly sheep	Cropping Mainly cereals	Pigs and poultry
275–449	69	80	74	40	64
450–599	56	59	—	42	—
600–1,199	36	45	48	21	44
1,200–1,799	18	21	40	12	24
1,800–	6	10	19	4	14
ALL SIZES	30	46	50	19	34
Unpaid family labour as a percentage of total labour costs					
ALL SIZES	7	8	22	3	8

* excluding horticulture

Source: Farm Management Survey.

wages and employment in England and Wales showed that, of the whole-time regular labour force in 1965, 70,700 (22 per cent.) were relatives, 64,400 of whom were unpaid, or more strictly 'non-hired', since they had no contract of service with the farmer, frequently their father (46,800 and 7,400 were sons and daughters respectively); unfortunately, the inquiry did not provide information about the distribution of such labour throughout England and Wales. For Scotland the December census shows that 10,312 of 46,386 regular workers (22 per cent.) were members of the occupier's family; but the proportion varies considerably throughout the country, as it probably does in England and Wales. In East Central Scotland, which includes the main arable areas, such family labour accounted for only 14 per cent. of regular hired labour, compared with 31 per cent. in the crofting counties, where most women employees were members of the occupier's family.

The Farm Management Survey of England and Wales throws further light on the contribution of relatives and of the farmer and his wife, although this sample includes a disproportionate share of large businesses and, probably, of progressive farmers. Unpaid family labour was relatively, but not absolutely, more important in the smaller businesses, but varied with farming type; a selection of average figures is given in Table 10. Not surprisingly, the relative contribution of the labour of the farmer and his wife (not separately distinguished) also declined with size of business, although absolutely it fell only slowly. It also varied with type of farm and was about twice as high on specialist dairy farms as on cereal farms, both on average and in most size groups.

LABOUR AVAILABILITY & DEMAND

The problems of identifying both farmers and farm workers and of equating the different kinds of full-, part- and spare-time employment complicates any regional comparisons of the relative importance of farmers and farm workers. Figure 6, computed from the 1966 population census, shows the proportion of all those employed in agriculture who are farmers and managers. It thus indicates broadly the ratio of farmers to farm workers and confirms that farmers and managers greatly outnumber farm workers in western districts, where farms are often small livestock-rearing or dairy farms, heavily dependent on family labour, and that the proportions are reversed in eastern counties, where farms are larger and cropping and horticulture predominate. The contrasts are greatest between Wales, where farmers outnumber farm workers by about three to one, and East Anglia, where the reverse is true. They are least in Scotland, where upland farms are generally much larger than in Wales, the paid labour force is proportionately larger, and the Highland counties make unsatisfactory statistical units.

The composition of the labour force is relevant to any discussion of regional differences in the availability of labour, for the different classes of labour must be equated in some way.

A scale of half a labour unit for each part-time worker and one-third for each casual worker is probably fairly near the truth, although the wages survey, with average earnings as a guide, used factors of 0·58 for regular full-time youths, 0·66 for regular full-time women and 0·47 for part-time casual workers of both sexes, obtaining a total of 350,900 man-equivalents for England and Wales, compared with a total labour force of 450,515 workers of all kinds. Because of the predominance of full-time regular male workers, mapping the results reveals a broadly similar pattern of distribution.

Unfortunately one cannot compute from the agricultural censuses any very meaningful figures for total labour, because of the difficulty of assessing the contribution of the occupier and his wife. With no suitable data, an alternative approach is to consider the distribution of estimated labour requirements, computed from census data on numbers by the application of standard man-days (Chapter 14). As Figure 21 shows, the pattern of labour requirements is very similar to that of regular labour (Figure 6). This approach has many limitations, for not only does the labour force vary in its age and sex composition throughout the country, but one should not expect even theoretical labour requirements to be the same everywhere, given differences in terrain, field and farm size, and levels and standards of mechanisation (especially in harvesting equipment). In practice, this latter difficulty is less serious than appears at first sight, since farming in the west and north is predominantly pastoral and the contribution of labour requirements for crops (the major source of error) represents only a small part of total requirements. In any case, any attempt to employ different standards for different parts of the country itself poses many problems, for it gives rise to discontinuities with no basis in reality. It would be valuable if such estimates of labour requirements could be matched against availability of labour to identify areas with apparent 'surpluses' and 'deficiencies'. Unfortunately, the difficulties of producing a map of total labour and the limitations of standard man-days make any regional analysis impossible, although a national comparison has been attempted by G. P. Hirsch.

The differences in labour requirements shown in Figure 21 result from several factors, of which farm size and farm type are the two most important. Small farms are generally more intensively farmed than large, through higher inputs of labour, fertiliser and purchased feed per unit area. Type of farming is also relevant and is related to farm size; for intensive enterprises such as dairying are more commonly found on small farms and extensive enterprises such as cereal growing on large. Figure 21 therefore gives some indication of the intensiveness of farming.

Demand for labour also varies throughout the year. It is more constant in branches of livestock farming, where animals require daily attention, than in crop production, where harvest represents a peak, although the growing of several crops, maturing at different times, helps to spread demand. The range of variation differs throughout the country, depending on the enterprises undertaken, the level of mechanisation and the number of enterprises on the holding. Thus, averages for regular whole-time workers in sample areas in England and Wales for 1946–65 show that, whereas cowmen worked 0·3 hours below the annual average in October–December and 1·1 hours above in July–September, the range for tractormen was approximately three times as large, averaging 1·0 hours below in winter and 3·1 hours above in summer; monthly and seasonal variations would, of course, be greater. This unevenness of labour requirement can be somewhat reduced by doing non-essential tasks in slack periods and by employing casual labour at peak periods; traditionally maintenance, hedging and ditching and the like have been undertaken in winter. In England and Wales, seasonal labour to meet peak demands appears to be increasing relative to other labour, suggesting a greater reliance on part-time labour, for part-time and seasonal workers were 30 per cent. of the total labour force in 1965 compared with 21 per cent. in 1949. In Scotland, however, where the proportion of part-time and seasonal workers is lower, their number fell more rapidly between 1951 and 1966 than that of full-time workers; the lesser importance of cropping, and especially horticulture, is probably a major reason. Mechanisation is reducing seasonality in some farm operations by permitting greater flexibility in farm operations and by eliminating the need for a large labour force at harvest time; some forms of mechanisation, however, have accentuated peaks by concentrating tasks into a shorter period. Unfortunately, full advantage of the elimination of seasonal labour peaks cannot be taken if others remain: sugar beet harvesting has been mechanised, but singling of beet still requires much labour. On small farms, with only one or two men employed, it may not be possible to use effectively the labour saved, since the total labour force cannot be reduced. Contract labour and machinery offer another way of eliminating seasonal contrasts in labour requirements, but they are still relatively unimportant; in Scotland, for example, the role of the contractor is mainly to provide expensive machinery which is used for only a small part of the year.

With no detailed analysis available, it is impossible to assess the significance of these differences in the availability and composition of farm labour, but several general observations can be made. Firstly, labour is a major input, accounting for 31 per cent. of all inputs other than seeds, feeding stuffs and purchased livestock. Information on how far this proportion varies regionally is inadequate, although there is some variation with type of farm. In England and Wales, according to the 1967 Farm Management Survey, the proportion of labour inputs ranged from 24 per cent. on cropping (mostly cereals) farms to 40 per cent. on horticultural holdings; in Scotland, where proportions are generally higher, it ranged from 30 per cent. on arable rearing and feeding farms to 46 per cent. on hill sheep farms. There are, of course, major differences in labour inputs per acre, ranging from £0·4 on hill sheep farms in Scotland, to £71·9 on horticultural holdings in England and Wales.

The availability and requirements of labour are also important because the 'chunkiness' of labour supply makes it difficult to balance them, especially on small farms, where a reduction from 2 to 1 in the regular labour force may represent a 30–40 per cent. reduction in total labour available. On larger farms a better fit may be achieved and benefit gained from specialisation of workers. However, the size of the labour force on the larger farms is declining and casual labour is becoming more difficult to find, so that the problem of fitting labour to land and other resources will become more widespread. The availability or lack of labour in general and of particular classes of labour also appears to impede agricultural change, although this is probably less true now than formerly, since sophisticated machinery often reduces the demand for labour dramatically, as in hop harvesting. The survival and expansion of horticulture and intensive crop production in East Anglia owes much to a tradition of regular part-time work among women, and piece rates and part-time working combine to produce high levels of family earnings and so to maintain an adequate labour force. Attempts to introduce horticulture to physically suitable areas which lack an adequate labour force or any such tradition face difficulties; carrot growing in the Fenland could probably not have expanded as rapidly as it did without a supply of experienced casual labour. Conversely, the absence of labour has probably contributed to the running down of much hill land. The nature of the farm work itself may be a factor, and dislike of a seven-day week has certainly influenced decisions to give up dairying.

Like other aspects of British farming, the labour situation has been changing rapidly, although data are not strictly comparable. In the mid-nineteenth century, when rural and urban populations were roughly equal, there were almost one and a half million farm workers in England and Wales on an agricultural area rather larger than at present, but depression in arable farming in the late nineteenth century and the disparity in urban and rural wage rates induced a steady migration of farm workers, and by 1951 there were under half this number; Scotland had similar trends. From 1921 onwards one can study trends in numbers of farm workers from the agricultural censuses. Continued depression in agriculture encouraged migration and, apart from a temporary rise immediately after the Second World War, the number of agricultural workers in Great Britain declined steadily, from 996,000 in 1921 to 451,000 in 1967, although the rate has been faster since 1950, when agriculture has been more prosperous than for some considerable time. Movement out fluctuates in the short run with national prosperity and the rate is higher in semi-urban areas; but the greatest volume of migration has come from the major regions of arable farming, such as East Anglia.

In contrast the numbers of farmers have changed little until recently. There is a strong element of continuity in farming and high proportions of farmers are the sons or relatives of farmers; farming is also valued for its independence and for the variety of work. Although there are farmers who earn less than farm labourers, they remain in farming; the farmstead is their home and they know no other occupation. Nevertheless, numbers are falling at the rate of 2–3,000 a year in England and Wales alone; the increasing competitiveness of farming and losses to urban development are important factors, and farms are often amalgamated on the death or retirement of the existing occupier. Nevertheless, farmers represent an increasing proportion of a diminishing labour force.

MACHINERY

With this decline in the labour force has gone a sharp rise in the mechanisation of British farming, particularly in the past three decades. Machinery is now almost as important an input as labour, accounting for 26 per cent. of all inputs, and in some parts of the country and in some types of farming it is more so; for example, in the cropping (mostly cereals) farms in the Farm Management Survey in England and Wales, most of them in east and south-east England, machinery costs in 1967 represented 30 per cent. and labour costs 24 per cent. of inputs other than seeds and feeding-stuffs.

No systematic collection of data on machinery was attempted before 1942. In England and Wales they are collected from a one-third sample of holdings, so that figures for the smaller counties are suspect, and not all items are enumerated each year; in Scotland a total count is taken every three years. Numbers of machines alone can be misleading, for there are considerable differences in the size and efficiency of different implements, and stability or even decline in numbers may conceal trends towards more powerful and efficient machines which can cover a bigger area in a given time. It is estimated that the power of tractors sold by dealers in the past two decades has doubled, that horse-power increased by four-fifths from 1957 to 1967, and that these larger machines required a quarter fewer drivers to do the same work. Furthermore, older machines (especially tractors) may be kept, perhaps as an insurance by smaller farmers, and are recorded in the returns even when they are little used, a situation thought likely in a sample survey in Scotland in 1967 which found that 21 per cent. of all tractors were over ten years old. Machines may thus be used for different proportions of the possible maximum but cannot be weighted accordingly, as was done with part-time and casual labour. Most farm machinery, too, is mobile and may be transferred between farms, especially where machines are highly specialised and used for only a small part of the year, or where there are many linked farms. There are no systematic data about inter-farm transfers, although the trends towards larger farms and more specialised machines suggest that they will become more common. Such movements between widely-separated farms are more likely for machines than for farm workers, who are tied by homes and families, although itinerant workers were once common, e.g. for the hay harvest. Like labour, machinery is available on contract. Contract hire is more important on smaller farms; on farms in the Farm Management Survey for England and Wales contract hire accounted for 16 per cent. of machinery and power costs on farms with under 450 man-days, compared with 7 per cent. on farms with 1,800 standard man-days and over. On a sub-sample of such farms three-quarters of farmers regularly used contractors, mainly for muck-spreading (33 per cent. of all operations), combining (18 per cent.) and beet pulling (12 per cent.). The machinery census in Scotland suggests that contracting is much more common in respect of the more specialised machines.

LEVELS OF MECHANISATION

All this suggests that conclusions from census figures about levels of mechanisation throughout the country must be treated with caution. No index of mechanisation is possible, for apart from monetary yardsticks, such as machinery costs or capital invested in machinery, which are not generally available on a regional basis, the great variety of machinery recorded in the census cannot be converted to some common denominator. The tractor is the most widespread machine, but there is a wide range of tillage and harvesting equipment, as well as stationary

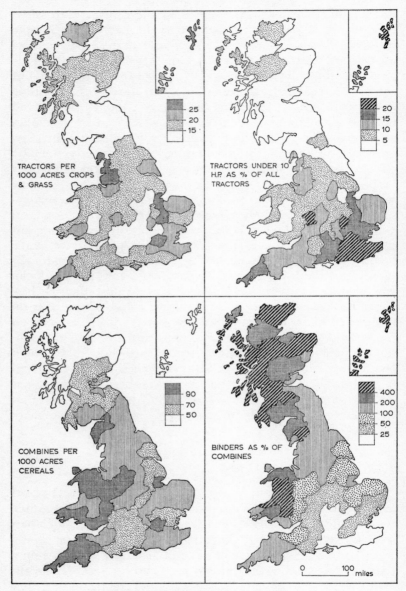

Figure 7. Mechanisation in agriculture
Source: machinery censuses, England and Wales 1965 (one-third sample) and Scotland 1964

machinery such as drying plant, milking machines and corn mills, each with a somewhat different distribution. The number of tractors is probably the best general index, although there are considerable regional differences in kinds of tractors and probably in levels of use. Conclusions will therefore depend on the basis of comparison. Absolutely, tractors are most numerous in eastern England, but numbers in relation to the acreage of crops and grass show surprisingly small regional differences, western counties recording quite high values and even Sutherland having more than two tractors per hundred acres (Figure 7). Partly, this reflects the wide range of tasks tractors perform, whether pulling tillage implements or grass mowers, operating stationary machinery, or lifting and transporting loads; some high values in upland counties undoubtedly reflect the importance of the tractor for personal transport over rough ground. The proportion of small tractors (under 10 h.p.), recorded in the census is highest in areas of intensive cultivation, notably in east and south-east England (Figure 7); however, county figures of total horse power per hundred acres of crops and grass, if available, would probably show that eastern counties had, on average, a marked advantage in available power over those farther west.

Other machines show a similar contrast between absolute and relative densities. If harvesting or other specialised equipment is related to the acreage of the crop being grown or harvested, western districts, with generally small acreages, often record the highest values; thus, when combine harvesters are related to the acreage under cereals, the leading counties, with values exceeding 9 combines per hundred acres, are in Wales and south-west England, with 16 in Carmarthen, where cereals account for only 4 per cent. of the crops and grass acreage, compared with 5 in the main cereal-growing counties of southern England (Figure 7). The reasons are partly that the machines in western counties are often smaller, and partly that combines are used less efficiently where cereal acreages are small. The contrast between Wales and south-west England on the one hand and northern Scotland on the other arises partly because the cereal crop in these northern counties is mostly oats, which are often harvested green and are less suited to combining than other cereals, and partly because the adoption of combines has spread outwards from southern England. Differences in the importance of contracting may also be a factor. The different forces at work also produce regional variations in the relative importance of those kinds of equipment; reaper binders (which are declining in numbers and are used largely for harvesting small acreages of oats) were more numerous than combine harvesters (numbers of which are rising) in all counties of the north and west, with ratios exceeding ten to one in north-west Scotland, while combine harvesters were four times as common as binders in southern counties (Figure 7).

Mechanisation varies in response to many factors. The nature of the terrain is relevant, although steepness of slope does not appear to impose any practical limitation in areas where reasonable crops can be grown. Soil type is important, with areas of heavy soil requiring more power and greater speeds of operation. Small farms and fields act as deterrents to the use of machinery and provide incentives to their enlargement; the number of tractors per thousand acres is much higher in north-east than in south-east Scotland, largely because of differences in farm size, and those in the north-east are less fully utilised. Conversely, the larger the farm, the greater the acreage of tillage per tractor; in a sample survey in 1952, there were 37 acres of tillage per tractor on farms with only one tractor and 71 on those with six, although there are obviously variations between farms of different types. Types of farming also affect levels of mechanisation as a whole, although for any one type of farm the differences between the proportion of inputs attributable to machinery and fuel vary surprisingly little between regions. On Farm Management Survey farms in England and Wales in 1967, machinery costs were proportionally lowest on horticultural holdings, where they accounted for only 16 per cent. of inputs other than feeding stuffs and seeds, and ranged between 27 per cent. and 34 per cent. on farms of other types; in Scotland they were lowest on hill sheep farms (17 per cent.), but otherwise varied little at 27–28 per cent. As with labour costs, the expenditure on machinery and fuel per acre varies widely, from £30·1 on horticultural holdings, to £11·1 on general cropping farms and to under £0·2 on hill sheep farms in Scotland.

EFFECTS & TRENDS

The recent mechanisation of agriculture has several consequences for the agricultural geography of Great Britain. First, the replacement of horses by tractors has released land formerly used to provide them with food; it required approximately 4 acres of crops and grass to support each horse, so that some 3–4 million acres have become available for food production in the past 60 years, more than half of them since 1939. The counties with most farm horses were generally those in the lowlands where losses of agricultural land to urban and industrial development have been greatest, and the release of land has largely offset the consequences of such losses.

Secondly, mechanisation has affected assessments of land quality. Mechanisation has both made it easier to cultivate heavy land (although the risks of failure remain greater than on most types of land) and also placed a premium on gently-undulating well-drained land which can be laid out in large fields and cultivated most easily. More heavy land is now cultivated than in the 1930s, even allowing for the general increase in the proportion of land under tillage, although the acreage is much smaller than at the war-time peak. However, it is in areas such as the Berkshire and Hampshire Downs that mechanisation has most affected land values (see Chapter 3).

Thirdly, mechanised farming influences the size of fields and farms. Speed of operation is reduced on small fields by more frequent turns and operating costs are therefore higher. So the desire to simplify mechanised farming and reduce its costs are important reasons for the widespread enlargement of fields, although the abandonment of livestock enterprises on arable farms has also made hedges less necessary. Similar forces are at work in farm enlargement: a 100 h.p. tractor, costing £5,000, requires a large acreage to justify its use. Because there is often a similar variety of work on small and large farms, the per acre capital costs of machinery are higher on small farms; investment in machinery on a farm of 50 acres may be £6–£9 per acre to produce the same level of mechanisation as £1–£2 per acre will provide on an 800 acre farm. Mechanisation becomes more sophisticated with larger farms and the proportion of machinery costs represented by depreciation tends to rise with size of farm. The relationship between farm size and mechanisation, however, is far from simple and on some types of farm in the Farm Management Survey machinery costs per acre were higher on the larger farms, possibly because of lower levels of mechanisation on small farms and of lower depreciation costs on second-hand machinery; smaller farms also use contractors more. Mechanisation is also often a status symbol and, because machinery comes in large units in relation to the size of farm, it is difficult to get a good fit between needs and mechanised equipment, especially on small farms where machinery may be under-used. The consequences of this are, however, lessened by the fact that most small farms are located where the emphasis is on grassland and grazing livestock, which lend themselves less readily to mechanisation.

Mechanisation saves labour and makes it more productive, and the lower level of mechanisation in grassland farming compared with arable farming accentuates regional differences; for the pressure to enlarge farms and fields is strongest in eastern counties, where the growing of cash crops is increasingly concentrated. The cost of specialised equipment also contributes to greater simplification of farming systems.

The replacement of men by machines powered by diesel oil, petrol or electricity is more recent and has happened more rapidly than the decline of the labour force. The introduction of tractors began before the First World War; by 1925 there were under 5,000, and in 1939 approximately 55,000. Thereafter farms were rapidly mechanised and by 1945 there were over 200,000. A maximum of over 480,000 was reached in the mid-1950s and there has since been a small decline, largely because of the increase in larger, more powerful machines; in Scotland, however, numbers continue to rise.

These facts illustrate general tendencies affecting most kinds of mechanisation and having both temporal and spatial components. In time, a new type of machinery is introduced, increases in popularity and numbers rise sharply. Then, as most farmers have acquired one, the rate of increase declines and numbers reach a peak, often to decline as new and more powerful versions are developed, or some new type of machine is introduced, or because the equipment is used more efficiently. Different kinds of mechanised equipment are at various stages in this sequence; reaper-binders reached their peak in 1950, at over 150,000, but were little more than a third as many by the late 1960s. Combine harvesters, with under 1,000 in 1942, have not yet reached their peak. Numbers increased eightfold between 1948 and 1958, when there were over 40,000, and have since increased more slowly to over 60,000; but the number of tractor-drawn combines is declining. The number of potato spinners reached a peak in 1954 and has since declined by a third; but the number of complete potato harvesters is still rising sharply and trebled between 1960 and 1965. In stationary power on the farm, the petrol or oil engine is being replaced by the electric motor. Numbers of milking machines, the only important application of power to livestock farming, reached their peak in England and Wales about 1960, but are still increasing in Scotland.

The rate and scale of mechanisation have not been the same everywhere. Equipment has generally been adopted first in one part of the country, often where the enterprise for which it was designed is well-established, and has spread outwards; the process has been studied as part of the general question of diffusion of agricultural innovations by G. E. Jones, who compares actual and theoretical patterns of diffusion. Lack of comparable data restricts analysis to numbers of machines, and changes in power and size have to be ignored. Nevertheless, examining the spread of equipment in this way is useful. For example, combine harvesters first became prominent in the corn-growing counties of southern England, where Wiltshire and Gloucestershire were the first to achieve 1 combine per 1,000 acres of cereals in 1946; by 1950 most of southern England except the south-west had achieved this ratio, eastern Scotland reached it by 1952, most of northern England except the north-west by 1954, while in most of Wales and Scotland it was not achieved until the late 1950s. The adoption of corn driers shows a similar northward extension, although Northumberland and Nottinghamshire reached a level of 1 drier per 1,000 acres of cereals at the same time as southern England; the last counties to achieve this level are the wettest counties of Scotland and Wales, where only a small acreage of cereals is grown.

Apart from the development of more powerful and sophisticated machines, a current major trend is the formation of machinery syndicates for the co-operative ownership of equipment. The movement appears to have begun in Hampshire, and between 1955 and 1960 some 65 syndicates were formed in sixteen counties; by 1967 there were over a thousand in 47 of the 66 counties of England and Wales, possessing equipment worth perhaps £2 million. Such syndicates must clearly affect the interpretation of the levels of mechanisation throughout the country.

The contrary trends in farm labour and farm machinery are affecting the relative importance of these two inputs. Figures for the United Kingdom as a whole show that labour, apart from the Second World War and its immediate aftermath, represented a declining proportion of inputs, falling from 27 per cent. in 1938–9 to 20 per cent. in 1966–7, while the contribution of machines rose from 6 per cent. to 16 per cent. The importance of machinery will undoubtedly increase as farms grow larger, as mechanisation is extended to livestock farming, as industrialised farming becomes more widespread, and as the labour force continues to decline.

FURTHER READING

D. J. Ansell and A. K. Giles, *The Farmer and His Time*, Misc. Study No. 46 (1969), Dept. of Agricultural Econ., Univ. of Reading

J. Ashton and B. E. Cracknell, 'Agricultural holdings and farm business structure in England and Wales', *J. Agric. Econ.* 14 (1960–1)

G. E. Jones, 'The diffusion of agricultural innovations', *J. Agric. Econ.* 15 (1962–3)

I. M. L. Robertson, 'The occupational structure and distribution of rural population in England and Wales', *Scot. Geogr. Mag.* 77 (1961)

P. M. Scola, 'Scotland's farms and farmers', *Scot. Agric. Econ.* 11 (1961)

W. M. Williams, 'The social study of family farming', *Geogr. J.* 129 (1963)

CHAPTER 6

Markets and Marketing

A market may mean a place where buyers and sellers, or their agents, regularly meet, or it may, as in this chapter, embrace the destinations, intermediate and final, of agricultural produce. Markets might be expected to figure prominently in any account of the agricultural geography of a country. In Von Thünen's Isolated State, the distance of a farm from the single central market dictated the kind of agriculture practised, and theoretical investigations of the location of agricultural activity have given first place to the relationship between producer and market. Yet most writings on the agricultural geography of Great Britain either do not mention markets or assume relationships which they do not demonstrate. With the notable exception of D. K. Britton's *Cereals in the United Kingdom*, studies by agricultural economists are not much more helpful on the pattern of produce flows and their costs, for farm accounts do not normally distinguish costs of transport which are borne by producers, and ignore those beyond the point of first sale. Thus, while there is considerable information about market structure and methods of marketing, little exists about the pattern of movement of produce from farm to consumer or about the costs of such movement. Nor, despite various official enquiries, such as the Verdon-Smith report on fatstock and meat marketing and the Runciman report on horticultural marketing, is much available from official sources; indeed, such reports mention the relationships between markets and producing areas only in passing. This chapter will examine some of these relationships, particularly the varied ways in which produce reaches its markets and in which the effects of distance are modified. The consequences for the location of agricultural production will be discussed more fully in subsequent chapters.

GENERAL CONSIDERATIONS

The marketing of agricultural produce differs from that of manufactured articles in several ways; largely for meteorological reasons, which are outside the farmer's control, the supply of produce is variable, sometimes highly so, but the demand for agricultural produce is relatively stable, so that problems of the disposal of variable surpluses arise. Much produce is perishable and cannot be stored for long (although this situation is changing) and production is by a large number of small independent producers, who cannot wait long for the sale of their produce. Buyers, by contrast, are increasingly organised in large units and, to offset the advantages this brings in dealing with many small producers, there have arisen various associations of producers, whether statutory marketing boards, co-operatives or business groups. For these reasons, and because agricultural products differ greatly in regularity of supply, perishability and other respects, very varied marketing channels have evolved; and because marketing has developed over a long period, traditional outlets and attitudes are often important.

The characteristics of agricultural marketing thus provide another reason why markets have received scant attention. In Von Thünen's Isolated State, there was a single market for all produce in a central city, but in a modern economy market relationships are much more complex and are difficult to identify. Each producer may have several marketing channels open to him and, given the small size of Great Britain and its complex urban geography, there are many possible markets for each commodity, unevenly distributed throughout the country. Furthermore, while the non-farm population (80–90 per cent. of which is urban) represents the ultimate destination of nearly all produce for human consumption which leaves the farm, in most transactions there are intermediate stages between producer and consumer, and produce does not necessarily move to the nearest town or, if it does, move by the shortest route. The simple linear relationship between distance and cost assumed in elementary location theory thus bears little relationship to reality.

Given the complexity of agricultural marketing, the general-purpose character of much traffic to and from farms and the existence of joint and return loads, it is often difficult to compute costs, and farmers, merchants and other buyers who provide transport may have little idea of its true cost; the United Kingdom cereal survey found that merchants often regarded transport as a service to farmers. Only where transport is hired or where single-purpose vehicles (such as milk tankers) are used are costs generally known. Furthermore, costs are modified in various ways by the pricing policies of transport firms and wholesalers, by differences in the volume of traffic and by government intervention, particularly through statutory marketing boards.

The question 'What are the markets for various products from different producing areas?' thus admits no simple answer. A small volume of transactions is conducted at the farm gate or at roadside stalls, for example, the sale of 'free-range' eggs, fruit or vegetables. Some produce, such as fodder crops and store livestock, is sold direct to other farmers without passing through any intermediate market; such farms may be situated nearby, but others are distant, as with the movement of hay from Stirlingshire to the Hebrides. Alternatively, store livestock destined for other farms may be bought by itinerant dealers or may pass through a store market and be sold by auction. A little produce is sold in urban markets by producer-retailers, although these are much less important than formerly, and other produce may be sold direct to retailers. Nevertheless, much farm produce passes through the hands of wholesalers or dealers, perhaps going first to some intermediate market to be bulked for transmission to retailers or processors. Still other produce is sent direct to processing factories, located either in the producing areas, as with quick-freezing plant, or in a town some distance away, as with grain for milling. The produce bought by such large wholesalers and processors ultimately reaches a much more widely-dispersed market, which may be national, as with quick-frozen vegetables.

The producer normally pays for transport to the point of first sale, although what he is charged may not represent the true cost, as the example of milk will show. In theory, producers pay the cost of transport to final markets indirectly through differences in the price they receive for their produce, the more favourably located producers receiving a higher price than those remoter from their markets; but evidence is scanty and any such differences may well be hidden by the influence of other factors. One of the few studies of transport costs, which examined their effects on farmers in the crofting counties, found the evidence inconclusive. For many products, such indirect costs are likely to be equalised among all producers, so that, in effect (though rarely explicitly) the more favourably-located producers subsidise the cost of transport from farms which lie further from their markets.

Various transport media are used, but like inland freight in general (61 per cent. of which moved by road in 1966) most agricultural produce goes by road and the share of rail transport has fallen sharply in the post-war period; for example, the number of cattle and sheep transported by rail fell from 1,750,000 and 3,500,000 head respectively in 1950 to 700,000 and 260,000 in 1962. Farming is so dispersed that the flexibility of road transport has obvious advantages, especially over short distances, and rail transport has been further handicapped by the closing of many country depots. Full discussion of transport pricing policy is not appropriate here, but certain features may be noted. Transport costs are rarely directly proportional to distance, and rates normally fall with distance. In the enquiry into transport costs in the Highlands, there was evidence that this was so, and such tapering is common in rail transport: in 1956 a crate of broccoli cost 2s. 3d. to send 100 miles from Penzance to Bristol and 3s. 5d. to send 340 miles to Liverpool. Transport rates may also depend on zoning rather than on the actual distance travelled; milk marketing provides an excellent example of this, but some purchasers of livestock deadweight adopt similar policies. Not uncommonly co-operatives and other large organisations adopt a flat rate for all produce irrespective of distance, based on past experience. In such circumstances the market is irrelevant to the individual producer (who has lost his choice of destination) and their importance as a factor in locating agricultural production is reduced or even eliminated.

Other considerations affect the cost of transport. Handling costs and size of consignment are important and unit costs are generally lower for full than for part loads; in Bedfordshire, for example, in the mid-1950s L. G. Bennett showed that some farmers were purchasing produce from others to make up full loads to distant markets. Similarly, M. Chisholm found that differences in costs of milk collection were associated with the volume of milk output per unit area and varied, *inter alia*, with the extent to which the full capacity of vehicles was used. Again, F.M.C. Ltd. (formerly the Fatstock Marketing Corporation) is said to have reduced costs of transport by a programmed collection to ensure full loads. The availability of return loads, locally or regionally, also reduces the costs of transport, especially by road. If transport bringing feeding stuffs, fertilisers and other inputs to farms can be used to carry produce, at least to its first destination, economies are effected. Similarly, the costs of inter-regional transfers are reduced if a return load, not necessarily of agricultural commodities, can be obtained; for example, there is much two-way traffic in coal from, and cereals to, South Wales.

Cost is only one aspect of the movement of produce determining choice of markets; speedy access is essential for perishable products, although technological developments are reducing its significance, as with the transport of milk, where treatment has gradually extended the possible range for fresh milk and where the development of long-keeping milk will have considerable effects on marketing in the future if costs can be reduced. Storage is also increasingly available on farms (as in grain marketing) or at intermediate points (as with cold stores for apples). Nevertheless, in many instances, especially in horticulture, freshness is important; processing provides the best example of such relationships, for peas for quick-freezing must reach the factory within $1\frac{1}{2}$–2 hours of picking, and pea-growing for this market is located mainly within 15–20 miles of freezing plant. The weight and bulkiness of produce in relation to its value also influences the distance that it can be transported to market, and the production of bulky commodities tends to be located near markets; for example, most of the potatoes marketed in Leeds come from the West Riding. Yet, even with bulky crops, quality may enable a product to enter more distant markets; thus the high reputation of swedes from Devon and Somerset enables them to compete successfully in Lancashire markets with those from more favourably located areas.

Furthermore, where there are several major markets, the location of production may be at some intermediate point between them.

Links between producers and purchasers may also be important. A farmer may sell all his produce to a particular market or a particular merchant because he has established a good working relationship or because it is more convenient to do so. Alternatively, a farmer may have a contract tying him to a particular outlet. As the size of business of both retailers and processors increases, this becomes increasingly important; crops are grown on contract for a particular factory, with seed provided and husbandry practices prescribed. Pigs and potatoes may similarly be produced on contract for particular outlets. Vertical integration between production and retailing or processing may also occur. Co-operatives represent another type of relationship and, although co-operation for sale of produce is not as well developed in Great Britain as in other European countries, co-operation is being encouraged by a Central Council for Agricultural and Horticultural Co-operation. J. H. Kirk has estimated that co-operatives probably account for 15–20 per cent. of all sales, being most common for cereals, eggs, potatoes, wool and horticultural produce. There is also an increasing number of farming groups who sell produce for their members; these are generally enterprising farmers who agree to sell all or a specified percentage of their output through the group. Most farmers without such commitments (which cover an increasing share of produce marketed) have a wide choice of markets, which may well be determined as much by habit and by personal predelection as by any conscious consideration of cost. Going to market is still a social event and some farmers spend a surprisingly high proportion of their time attending different markets.

If the movement of produce off farms could be monitored by some remote sensing device and all movements plotted cumulatively on maps, the results would be a network of extraordinary complexity, with lines criss-crossing in every direction, although there would probably be dominant trends terminating at the major urban centres. Some produce, such as milk, is marketed daily, but most commodities are marketed seasonally, so that the network would vary throughout the year; it would also change from day to day as farmers switched from one market to another as local gluts and shortages arose and prices weakened or hardened accordingly.

Without such a device and any suitable data, no comprehensive picture of movements to market is possible. Nor can market demand be equated exactly with population, for dependence on imports varies throughout the country and there are regional differences in diet; for example, imported meat is mostly consumed in central and southern England, and the per capita consumption of beef and veal is 49 per cent. higher in Scotland than in south-west England, while that of lamb and pork is 40 and 30 per cent. lower respectively. However, some insight into the relationship between producing areas and markets is gained by briefly considering the marketing of six products (cereals, horticultural products, root crops, livestock, milk and eggs) which illustrate the variety of channels and the range of degree of control over outlets.

CEREALS

Cereals and root crops are the bulkiest products moving off British farms, each accounting for over 13 million tons; but, whereas the movement of cereals is wholly determined by the decisions of farmers, merchants and processors (except in so far as farmers take advantage of the Home Grown Cereals Authority's bonus scheme for forward contracts), potato marketing is modified by the operation of a statutory marketing board and sugar beet marketing controlled by a statutory monopoly, the British Sugar Corporation. Furthermore, whereas major problems of disposal arise from the variability of supply of main-crop potatoes (in which the United Kingdom is normally self-sufficient), and sugar beet has only one buyer that accepts all the crop from a specified acreage, large quantities of cereals are imported and there is thus a potential market (at appropriate prices) for all production. The marketing of cereals has evolved through the free operation of market forces and the price guarantees give growers an incentive to realise the best market price by being related to average, not actual, prices. More is known about the movement of cereals than of any other commodity, except milk, through the comprehensive inquiry by D. K. Britton for the Home Grown Cereals Authority, the body set up in 1965 to improve the marketing of cereals.

Grain is harvested over a period of some six weeks in July, August and September, and stocks are then run down as grain is consumed on farms or sold to millers, distillers, manufacturers of feeding stuffs and others. Grain marketing is therefore concerned with both storage and transport, and is conducted mainly in two stages, collection by merchants, who then sell to other buyers. Grain is grown on some 150,000 holdings, although not all growers sell grain and 21 per cent. of those in D. K. Britton's survey sold none in 1967. Most of the grain is grown in the grain-surplus areas of eastern England and Scotland, extending from Essex to the East Riding and from Roxburghshire to Fife. Some grain is sold immediately and the remainder stored on farms for varying periods. Since combined grain is generally too moist for safe storage, it must be treated if lengthy storage is required. Lack of proper storage facilities and of grain driers is a major problem, especially for small growers, since it implies that the crop must be sold soon after harvest, when both actual and guaranteed prices are lowest; in the sample survey, only two-fifths of growers stored grain in purpose-built stores and only 31 per cent. had grain driers (though the proportion was 95 per cent. among those with 300 acres or more of cereals).

Most grain is purchased from growers by some 1,800 merchants, of whom a third account for four-fifths of the turnover, but there are also direct sales to processors of grain and, on a smaller scale, to other farmers. Merchants are widely distributed, but are most numerous in the main grain-producing areas, as a map in the Britton report shows: there are 10 per 100,000 acres in Suffolk and 11 in Hertfordshire, compared with 2 in Carmarthenshire and 3 in Northumberland. Most growers are thus quite near at least one merchant and over half the merchants do most of their trade within a 20-mile radius; only among the larger merchants do more than half transact most of their trade at distances above 30 miles. Merchants seek out farmers with grain to sell and transport it to other buyers; most of them also have storage facilities, although storage conflicts with the need for rapid turnover and under a quarter have any considerable capacity, most of these being processors as well as merchants.

Contacts between farmers and merchants are often close and affect the choice of outlets, for most merchants also deal in other commodities, especially feeding stuffs, fertilisers and seeds, which they supply to farmers, and thus can keep transport costs low by fuller use of their vehicles. Only a minority are specialist grain merchants, who are most common in eastern counties; nearly half of these do not own transport, finding it more economical to hire it.

Some of the grain purchased by merchants is made into feeding stuffs by those who combine the role of merchant and manufacturer and who handle the bulk of the grain sold, but most is sold either locally or to more distant markets. The long distance movements are mainly to the major ports on both east and west coasts, where the large millers and compounders are located, to other concentrations of processors, such as the whisky distilleries of north-east Scotland, and to the grain-deficit areas of western counties (although little of their requirements can be supplied from the home crop). Maps in the Britton report show that the pattern of movements is highly complex. Most grain is moved by road, which is preferred for its flexibility, although a return load is desirable for distances of over 30 miles and essential for those of over 100. Fortunately, the multiple interests of most merchants facilitate such back-loads, with lorries carrying grain in one direction and feeding stuffs or fertilisers in the other, and perhaps half of all journeys have a paying load in both directions. Movement is less well balanced in the remoter areas, such as north Scotland and south-west England, which are thereby at a disadvantage. About a third of merchants sometimes use rail transport and a third of these only very occasionally. A minimum load of 100 tons is required for rail transport, but it is advantageous for large long-distance loads; a weekly train of up to 30 wagons travels from near March, in the Isle of Ely, to Dufftown in Banffshire, carrying barley and malt for the distilleries.

The marketing of grain illustrates the role of merchants in bringing together supplies from numerous producers in the quantities, often large, required by industrial buyers. The diversity, both of growers and of outlets, makes the pattern of marketing complex, but it is efficient, partly because trade between merchants provides flexibility in meeting market demands, and partly because the diverse activities of merchants spreads overheads and reduces transport costs. The principal weakness is the inadequacy of storage capacity both on farms and at merchants' premises. This diversity of activities also means that grain marketing cannot be evaluated in isolation.

HORTICULTURAL PRODUCE

Horticultural marketing is similarly complex and likewise almost entirely the product of supply and demand, although there is a strong historical legacy in the siting of the main markets, many of which, including Covent Garden, are now being redeveloped with government assistance. Yet horticultural marketing contrasts with cereal marketing, not only in diversity of products, but also in centralisation of marketing functions. Horticultural produce comprises a large range of crops differing greatly in perishability, seasonality and value, and in soil and climatic requirements; moreover, these products can in part be substituted for one another and suppliers are usually highly variable, being greatly influenced by the weather. Unlike cereals, most horticultural produce is consumed fresh and marketed mainly at relatively few large wholesale markets in the major cities, although there is some small-scale local trade, and direct sales to large buyers and travelling wholesalers are increasing. An estimated 70 per cent. by value of all home horticultural produce passes through about 30 primary markets. The Runciman Committee, reporting in 1957, recognised major primary markets in London, which were by far the biggest, with Covent Garden's turnover alone exceeding £70,000,000; in Birmingham, Glasgow, Liverpool and Manchester, each handling produce worth well over £10,000,000; and in Bristol, Cardiff, Edinburgh, Hull, Leeds, Newcastle, Nottingham, Sheffield and Southampton. These primary wholesale markets serve several functions; they assemble a wide variety of produce; they serve directly the retailers in the large cities which, with their hinterlands, contain perhaps two-thirds of the population; they supply secondary wholesalers, who supply retailers in other parts of the country with those products not grown locally; and they act as price regulators. The flow of produce is so variable that fairly local markets would suffer from gluts and shortages, with widely fluctuating prices, for much horticultural produce is perishable and cannot easily be stored. Such surpluses are more easily absorbed and deficiencies made good in big markets. Furthermore, the wide variety of produce retailers require can be assembled from growers in different parts of the country only in large markets, where demand and supply can be matched. Moreover, many primary wholesale markets, and all the largest (except Birmingham), are also ports, where imported produce, which both complements and competes with home-grown produce, is assembled.

Covent Garden exercises a dominant role in marketing, especially in respect of the scarcer crops and those grown out of season; in an investigation by L. G. Bennett in the mid-1950s, London was mentioned most frequently as a market by growers as far away as west Cornwall. This dominance arises partly from the size of its immediate market and partly because London is not only the leading port but is the focus of both road and rail communications, so that produce can sometimes be received from producers or dispatched to distant markets more easily than from intermediate points; furthermore, if a journey must be made to fetch supplies of choice fruits and vegetables, other produce may as well be bought at the same time. The Runciman Committee found that some secondary wholesalers in south and south-west England and in the east Midlands obtained produce more conveniently from Covent Garden than from other major markets because of the better transport and the wider range of commodities. There is consequently some reconsignment to other major markets. The pre-eminence of Covent Garden may also owe something to market connections: celery grown in the Isle of Ely is said to be marketed there because of connections between the growers and Chatteris men in the market. Such links certainly matter; secondary wholesalers in Wales and south England told the Committee that they received regular supplies of Brussels sprouts direct from the Evesham area by virtue of long-established connections.

Sales to primary and secondary wholesalers are not the only outlets. Some produce is sold direct on wayside stalls and

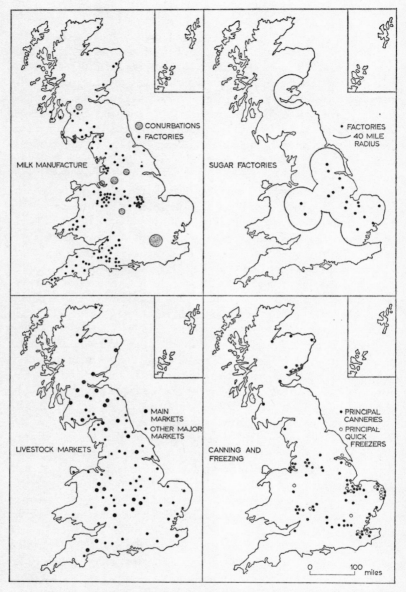

Figure 8. Markets for agricultural produce
Sources: milk marketing boards, British Sugar Corporation, farmers' unions and marketing officers, and the Fruit and Vegetable Canning and Quick Freezing Research Association

at local markets, and local retailers may also be supplied; travelling wholesalers are also increasingly making direct purchases from growers. More important are contract sales to major retailers who require quality crops marketed to their own specifications; about 15 per cent. of all produce is estimated to be handled by supermarkets and other retail chains. Such contracts are made with large growers, with growers' co-operatives and with country merchants, often substantial growers, who organise the production of a group of growers; an outstanding example of this is in cauliflower growing in Lincolnshire. Lastly, about a third of all produce goes to processors for canning, quick freezing or jam making, and the proportion is much higher for some crops, e.g. raspberries in Scotland. Canning and quick-freezing require a regular supply of high quality crops and much produce is grown on contract by farmers close to the factories, which are located in the main growing areas, often at ports, through which fish for quick freezing is also received (Figure 8). Such farmers are often substantial growers and the produce is either delivered to the factory or collected at the farm, so that transfer costs are minimised. With jam making, which absorbs much of the soft-fruit crop, regularity of supply is less essential, since fruit can be pulped and stored if necessary.

Before 1950 most horticultural produce went by rail, but perhaps 90 per cent. is now transported by road, for this gives greater flexibility both in destination (since produce can, if necessary, be switched en route) and in size of loads. Even in the mid-1950s only one of the four major areas investigated by L. G. Bennett, west Cornwall, which has poor road links, depended primarily on rail transport, and the Runciman Committee estimated that some 500,000–750,000 tons of produce reached Covent Garden by road, compared with 200,000 tons by rail. The shift to road transport has been helped by the advent of large lorries and by road improvements and the opening of motorways.

Markets vary with the type of crop. For bulky, low value crops, costs of transport are a major consideration; where, as generally happens, they can be widely grown, most of the produce is sold direct to local wholesalers and the proportion sold on commission at major markets is reported to have fallen in recent years; a sample of growers questioned by the Runciman Committee sold 69 per cent. of their cabbages direct, most of them to wholesalers, but 21 per cent. to retailers or by direct retail sale. By contrast, 64 per cent. of plums were also sold direct, processors taking 44 per cent. of the crop, and only 3 per cent. was sold retail or direct to retailers. Of the choicer vegetables, most is sold on commission at the major markets: with eating apples (Cox's) 84 per cent. were sold on commission (74 per cent. at major markets). A widely grown, but perishable commodity, such as lettuce, occupies an intermediate position: 63 per cent. of produce was sold direct (50 per cent. to wholesalers), but lettuce grown under glass is largely marketed through London markets. The proportions sold through different channels also vary from area to area; in L. G. Bennett's enquiry, for example, 88 per cent. of Lea Valley produce was sold through commission agents compared with only 54 per cent. from Wisbech, while retailers took 14 per cent. of the produce from Bedfordshire and only 5 per cent. of that from the Lea Valley.

The pattern of movement to ultimate markets is highly variable and even producers in East Lothian sometimes send produce to London. The complexities of movement can be illustrated at both pre- and post-market levels. L. G. Bennett instances the movements of a single lorry in Bedfordshire, which collected produce from 20 growers in small consignments and took it to London, where it was distributed to 8 different commission agents. The Runciman Report illustrates the complexity by a hypothetical example: a merchant in the Evesham area, who usually bought runner beans at a fixed price from a local producer and sold them to retailers within a 20-mile radius, was unable to sell all his produce one day because local demand was weak; he consigned some to a primary wholesaler in Birmingham for sale on commission. Prices in his market, which had been firm, also began to fall and, although the wholesaler sold most of his consignment to local retailers, he dispatched some to Glasgow following a telephone call from a wholesaler, since prices there were moving up. The Glasgow merchant calculated that he could buy beans in Birmingham and sell them at a profit after paying transport costs; a small part of his consignment from the Birmingham merchant went to a secondary wholesaler in Inverness, who in turn supplied a retailer in the Highlands. Proximity to market has little meaning in such a context.

Apart from the compulsory grading of produce sold wholesale, which is being gradually introduced and so far affects five commodities, the grower's choice of markets is limited by his accessibility to them in terms of perishability and transport costs. Acceptable standards of grading might permit more produce to be sold on description, rather than at primary markets, and so reduce waste and cross-haulage. Present arrangements, whereby variable quantities of often highly perishable commodities reach consumers from over 50,000 growers, largely through perhaps 1,500–2,000 wholesalers, do put buyers and sellers in touch reasonably well, although the rising proportion of direct sales to retailers and processors suggests that, for some growers and purchasers at least, there are advantages in bypassing the primary markets.

ROOT CROPS

The marketing of cash root crops will be treated more briefly. Less is known publicly about the movement of potatoes from farm to market, and that of sugar beet, while contrasting with the movement of other crops, is relatively simple.

Growers of one or more acres of potatoes for sale must register with the Potato Marketing Board, and are allocated a basic acreage of potatoes; in 1967 about 53,000 were registered. Potatoes suffer from the combination of very variable yield and a demand that is price inelastic. Imports of main crop potatoes are prohibited when the home crop is sufficient, but high yields create a surplus, which cannot be absorbed by price reductions, and so must be disposed of otherwise. Unlike the marketing boards for milk and eggs, the Potato Marketing Board does not normally buy potatoes, but exercises control by prescribing what proportion of their basic acreage growers may plant and, when necessary, the minimum size and standard of potatoes that may be marketed, measures of limited value since yield is the chief variable and years of glut are often years of big potatoes. The Board also administers the price guarantee for potatoes, supporting the market by buying potatoes if the average price seems unlikely to reach the guaranteed price.

Potatoes for human consumption can be marketed only through some 3,500 licensed merchants, who are widely distributed; there are also some 3,200 licensed grower-salesmen making direct sales to wholesalers, retailers and caterers. Most potatoes are sold for retailing in the main urban centres, but an increasing proportion is processed as crisps, dehydrated or canned potatoes, and these, together with prepacked potatoes, were estimated at 40 per cent. of sales in 1968. Such manufacturing plants are widely distributed (Figure 8). There are considerable regional differences both in per capita consumption of potatoes, which is high in northern England, and in preferences for different varieties; for example, the most popular varieties in Scotland are Redskin and Kerr's Pink, and varieties grown for use as seed potatoes in England, such as Majestic and King Edward, cannot easily be diverted to Scottish markets. The trade in potatoes is mainly over comparatively short distances and return loads are desirable for long journeys.

Potatoes surplus to human requirements at the end of the season are purchased by the Board for stock feeding or for processing for outlets other than human consumption. As much as one third of such potatoes may be sold back to growers for stock feed without leaving the farm; others are sold to merchants for resale as stockfeed.

Apart from its control over growers' acreages and the volume of potatoes marketed, and the restriction of outlets to licensed merchants and grower-merchants, the Board does not affect the pattern of movement of potatoes for human consumption, nor does it modify the cost of transporting them. For sugar beet, in contrast, the British Sugar Corporation is a monopoly buyer and there are only 18 outlets, the Corporation's factories, located mainly in eastern England (Figure 8). A grower wishing to sell beet must obtain a contract, which prescribes the maximum acreage from which he agrees to deliver beet to a specified factory. In England and Wales the grower who sends beet by road makes his own arrangements, using his own transport or negotiating the lowest rate he can with a haulier. Most of the beet is grown within 30 miles of a factory, but a grower sending beet by rail in consignments of at least 10 tons can be reimbursed for the difference between his payment per ton from his nearest railway station and the Corporation's prescribed rate per ton for 40 miles, although he pays the cost of delivery to the station. A distant producer is therefore no worse off than one who sends his beet from a station 40 miles from the factory; nevertheless, the quantity of beet sent long distances is small. The Corporation may divert beet to any other factory, but it then pays the difference in freight charge. In Scotland, with only the one factory at Cupar in Fife and sugar-beet growing more dispersed, the government and the Corporation jointly subsidise transport costs. All growers receive a basic payment per ton of clean beet (i.e. after soil, stones and trash have been removed) and an additional payment for every ton-mile up to 80 miles along the shortest route between the steading and the factory; in 1968, the basic payment was 11s. 6d. and the rate per ton-mile 3d. Most growers employ hauliers, but some use their own transport and many within ten miles of the factory deliver by tractor.

The effects of this policy are difficult to assess from published information. In England and Wales, only 1 per cent. is grown in counties lying further than 40 miles from a factory, but in Scotland 11 per cent. of the contracted acreage lay in such counties in 1968, though very little was grown at distances beyond 80 miles. Nevertheless, despite the transport subsidy, the sugar beet acreage in Scotland has declined and it has been decided to close the Cupar factory.

LIVESTOCK

The marketing of livestock and livestock products exhibits perhaps even more varied marketing arrangements, from the market in store stock, where only tradition, linkages between farms or between farmers and dealers, and lack of knowledge limit the free interplay of supply and demand, and where transport costs are similarly determined, to the market in milk, where all producers must sell to the milk marketing boards, and in eggs, where those selling to the Egg Marketing Board receive the same price for consignments of the same size irrespective of location.

Cattle, sheep and pigs may be marketed as stores or as fat stock, although many animals are now reared and fattened on the farms on which they were born. There is little comprehensive information about either the marketing or the movement of store stock, but L. V. McEwan's study of marketing of store stock in Scotland throws some light on the trade. It contains a large random element, but the main patterns of movement in Scotland are clear: a general move from upland to lowland farms, and of store cattle from Ireland via west coast ports. Although some cattle are sold direct, the main outlet is the auction mart. There were 163 store markets in Scotland (86 of them seasonal) so that farmers need not send stock great distances to markets: in the study of Highland transport noted earlier, the average distance in crofting counties was 19 miles for store cattle and 32 miles for store lambs, compared with 16 and 19 miles respectively in other Scottish counties. A farmer may send his stock to market in either his own or hired transport or may sell to dealers or to other farmers. Dealers buy a considerable amount of stock from farmers as well as buying and selling store stock at auction marts; in 1962 they purchased 41 per cent. of the store calves bought at such marts in Scotland, 17 per cent. of the older store cattle, 18 per cent. of the store sheep and lambs, but only 8 per cent. of the store pigs. Store markets vary considerably in importance: some, such as the Border lamb sales at St. Boswells, are of national importance, while many serve only a local community; they also vary in their frequency and regularity.

Trade in fatstock is better known, for they are eligible for price guarantees, related (like those for cereals) to average and not actual realised prices, and so providing an incentive to realise the best market price. The Verdon-Smith committee's report gives a comprehensive picture of the marketing of fatstock and meat in the early 1960s, but little information about the regional pattern of marketing. Since the outlet for pigs is more specialised than that for sheep and cattle, this will be discussed separately.

Fatstock are produced on many lowland farms, but relatively few farmers provide most of the supplies: in Scotland one twentieth of the farmers produce over half the supplies of cattle. Most sellers are individuals, but there are groups who assemble and sort livestock for their members and make contact with wholesalers and factories; with pigs and intensive beef the members of such groups may work under contract for an assured market.

Over two-thirds of fat cattle and almost two-thirds of fat sheep, (but only about a fifth of pigs), are marketed through livestock auctions. In 1963 there were 856 livestock markets in Great Britain, and no farmer is very far from one. Although a few markets were exclusively for fatstock, the great majority also handled stores. Many markets are small, and two-thirds of them handle rather less than a third of the trade; Figure 8 shows the more important. After purchase, fat livestock are sent for slaughter, mainly in the urban areas, and for transmission to retailers, either direct or via meat wholesalers.

The remainder of the fatstock marketed, i.e. 25 per cent. of cattle, 30 per cent. of sheep and 80 per cent of pigs, were bought deadweight; there were then 969 approved deadweight centres, mostly slaughterhouses, but including 190 bacon factories. Livestock sold ordinary deadweight (i.e. without grading) were normally sold by a farmer to a retail butcher or wholesaler at the slaughterhouse; sales deadweight and grade might be arranged by the field staff of large wholesalers who canvass for supplies in producing areas. Under this system a farmer arranges his own transport to the slaughterhouse or the wholesaler does this for him, often making a standard charge for his services.

For pigs the specialised requirements of the main buyers and the comparative regularity of supply have favoured the use of both formal and informal contracts between producers and buyers; under the latter a buyer may agree, for example, to take the whole output of a particular producer. Contract sales have been encouraged by the government and four-fifths of all pigs sold deadweight and grade were sold under long-term contracts at the time of this enquiry.

Buyers of fatstock are much more diverse than sellers, who are either farmers, groups of farmers or dealers. They include retail butchers, wholesalers, bacon curers and dealers. Retail butchers obtain their supplies from auctions, from dealers and direct from producers, generally purchasing from only a few producers. Wholesalers buy through all channels, but the bacon factories mainly buy direct from farms. The contribution of dealers is least well-known; they normally purchase on commission for butchers and the like. A considerable proportion of livestock sold is handled by F.M.C. Ltd. (established in 1954 under the sponsorship of the National Farmers Union, but now a public company). It procures livestock, generally direct from farmers, and handles the slaughter of all classes of stock; it is particularly prominent in the marketing of bacon pigs and, when the Committee reported, was handling almost two-thirds of the pigs sold deadweight and grade.

Most livestock, whether sold liveweight or deadweight, are thus killed in slaughterhouses, of which there were 3,177 in Great Britain in 1962, although the number has fallen considerably since; many in England and Wales are small and there are also some 40 to 50 places in the Highlands and Islands where animals are occasionally slaughtered. Most slaughtering is done near the main urban markets; for cattle, just over half were slaughtered in the consuming areas, compared with a quarter in the producing areas, the latter proportion having increased between 1956 and 1962.

Thus, although there is a general easterly movement of store stock from the uplands to the lowlands and from the west coast ports, and a movement of meat and of fatstock for slaughter to the main urban markets, the details of the pattern are highly complex. The auction markets, through which most of the fatstock pass, not only bring buyers and sellers into contact, but also sort stock into appropriate lots and secure economic loads for onward transmission; dealers fulfil a similar role. Since these centres are near producers, they may save something in transport costs, but there is little evidence on this. Nor can one prove or disprove the assertion that auction markets represent an unnecessary link in the marketing chain, involving additional handling and transport charges; the evidence on this point also is inconclusive, and there appears to be little difference between prices at auction marts and under systems of deadweight marketing. In any case, most fatstock markets are also store markets and part of their costs of operation are attributable to this function; if they closed, many smaller store markets would also have to close, which would raise transport costs to farmers. As recommended by the Verdon-Smith report, a Meat and

Livestock Commission was established in 1967. It is charged, *inter alia*, with improving the marketing of livestock and meat, though it is not empowered to trade in either; one cannot say yet what effect, if any, it will have on the pattern of movements.

MILK

Milk is a highly perishable commodity, consumed mainly as liquid milk by the urban population; supply and demand vary from day-to-day, necessitating transfers between one market and another, although these fluctuations are much smaller than in horticulture. In contrast, however, to livestock marketing, all milk leaving the farm must be sold to producer-controlled marketing boards and the maximum retail selling price of liquid milk is fixed by the Government, which provides a guaranteed price for a standard quantity. Any excess above this quantity is sold at considerably lower prices for manufacture into milk products, which compete with imports. So the function of manufacture is to absorb any milk that cannot be marketed as liquid milk. Milk factories consequently fulfil a different role from most factories processing agricultural produce, which aim to operate at full capacity; milk factories, on the other hand, act as a safety valve for the liquid milk market and production varies greatly throughout the year, with some creameries closing down in winter when supplies are lowest. In England and Wales, for example, milk used for manufacture varied from 42 per cent. of all milk sold in May to 23 per cent. in January. Factories were originally located in dairying areas some distance from the major urban markets and producing milk cheaply on grass. Encouraged by the milk marketing boards, they remained in these areas because the major milk surpluses originate there and because many factories serve as depots for distant markets, where supplies can be collected and switched to other markets or manufacturing as required (Figure 8); in England and Wales this concentration has become more marked as several factories near urban markets have been closed. With the creation of the milk marketing boards, the pattern of supply and demand as a whole can be more easily seen; moreover, rationalisation of milk transport during the Second World War has eliminated much costly cross haulage which had originated earlier, so that in this sense they created a more efficient market for milk.

Four milk marketing boards cover Great Britain: the Milk Marketing Board, by far the largest, has responsibility for England and Wales, where 90 per cent. of the milk sold off farms is produced, while the Aberdeen and District (1 per cent. of milk sold), the North of Scotland ($\frac{1}{2}$ per cent.) and the Scottish Milk Marketing (8 per cent.) Boards cover the northeast, the north, and the centre and south of Scotland respectively. These boards were established to strengthen producers vis-a-vis buyers, who were increasingly dominated by large dairy companies, but they also reinforced the trends (discussed more fully in Chapter 8) towards a wider dispersal of dairying and, more recently, a westward shift in the centre of gravity of the dairy industry; for these trends in part arise from the pricing and transport policies of the boards.

Although the price obtained by boards for milk for manufacturing is less than half that for liquid milk, producers are paid a uniform price, so that the individual dairy farmer is no longer concerned whether his milk is sold for liquid consumption or for manufacturing (although there are marked regional differences in the percentage of milk manufactured, ranging from 30 per cent. in England and Wales to 42 per cent. in Scotland, and from 5 per cent. in eastern England to 56 per cent. in south-west England). For producers as a whole, however, the more milk sold for manufacturing, the lower is the price they receive for all their milk. For example, in 1968, the net returns to the milk marketing boards for liquid milk was 50·30d. per gallon and for manufacturing 21·70d. per gallon, and the guaranteed price was reduced by 3·00d. per gallon by the sale of manufacturing supplies above the standard quantity.

Since the boards are each self-contained, their creation itself modified the pattern of movement of milk to market, notably of the supplies formerly sent from south-west Scotland to northeast England, which ceased in 1934 by agreement between the Scottish Milk Marketing Board and the Milk Marketing Board. A more general cause of changes in relationship between producing areas and markets has been the transport policies of the boards, which have lessened the importance of location, especially to individual producers; for, given uniform prices to consumers, apart from differences for special grades of milk, the only variable other than production costs is that of transport costs borne by the producer.

Producers pay transport costs only to the first destination, whether this is an urban dairy or a country depot, but transport rates are not proportional to distance and the way they are determined varies among the different boards. England and Wales are divided into eleven regions, in which producers receive almost the same gross price, the differential at maximum being only 0·5d., little over 1 per cent. of the average price. In each region producers pay a standard transport rate per gallon irrespective of their location; these regional charges bear some relation to distance travelled, the Far Western (south-west peninsula) rate, for example, being 1·35d. per gallon, and the south-eastern rate 0·66d. per gallon. The differences are greater than the differences in average collection costs from farms, which range from 0·92d. per gallon in the southern region to 1·39d. in the eastern region, and in some degree reflect the differences in proximity to major markets. Nevertheless, about a quarter of the milk sold for liquid consumption goes to such markets from depots, at an average cost of 0·55d. per gallon, ranging from nil in the south-east region and 0·06d. in the East Midlands to 1·45d. in south-west England and South Wales, and the cost of this secondary transport is not paid by producers. This system of charging weakens the pull of urban markets, especially London, which draws much of its supplies from southern and mid-western England.

Transport rates in Scotland are related more directly to distance travelled, except in Orkney and the islands of Argyllshire, where a flat rate is payable. Even so, the charge is not based on actual mileage travelled, but on location within zones of varying width measured from the main markets; furthermore, to safeguard producers in east Scotland, additional markets were defined, many of them small towns, so that such producers would pay only the lower rates. These rates taper with distance; the Scottish Milk Marketing Board, for example, has five rates, viz. up to 5 miles (1·75d.), 5 miles to 10 miles (2·00d.), 10 miles to 30 miles (2·25d.), 30 miles to 40 miles (2·50d.) and over 40 miles (2·75d.), and milk travelling over 40 miles pays little more than $1\frac{1}{2}$ times the rate for distance of five miles or less. The other milk marketing boards also apply tapering rates.

Most of the milk is moved from farms either by purchasers or by hauliers, although milk marketing board depots themselves transport nearly a quarter of the milk from farms. Five-sixths of the milk is collected in churns, except in the Scottish and Aberdeen and District Boards' areas, where two-thirds is collected by tanker; but all boards now pay premiums to encourage bulk collection of milk. The proportion so collected is rising sharply and milk can now be collected direct from farms over virtually any distance. Milk sold by producer-retailers, once characteristic of farming on the urban fringe, is now largely confined to small towns and villages and accounts for only 3 per cent. of the milk sold in England and Wales and 8 per cent. in Scotland. A further 1 per cent. is sold to farmhouse cheese-makers. Four large organisations, with 277 establishments, handle 58 per cent. of milk sold in England and Wales and such milk moves, in roughly equal proportions, to processing dairies in the urban areas and to depots located mainly in the principal producing areas; of the latter supplies, approximately half is manufactured and the remainder sent to urban areas, mainly London, for the liquid market. In Scotland, where little milk is transported far and depots in the English sense are unnecessary, 95 per cent. of the milk not despatched direct is manufactured. The processing dairies deliver milk to consumers or sell it to non-processing dairymen, although a little is manufactured.

Factories processing milk not required for the liquid market also show some regional specialisation by function. Those near the urban areas use any surpluses for cream (although south-west England also produces large quantities). Butter, the least remunerative outlet, is produced mainly in south-west England and South Wales, while cheese is widely made, notably in Dorset, Somerset, Wiltshire and in the West Midlands.

This general pattern changes greatly with the seasons and in detail from day to day, although few movements are as paradoxical as two examples quoted by J. D. W. McQueen, a daily train carrying milk from Aberdeen to Dumfries and milk moving in summer from Edinburgh to Kirkcudbright.

Yet even these illustrate the difficulties of achieving balance between supply and demand.

EGGS

In many ways shell eggs provide the most interesting example of the spatial relationship between producers and their markets, for these reach consumers through both a free market, in which large numbers of producers sell direct to consumers, and a regulated market in which all producers receive the same price (though prices can vary considerably because of bonuses for large consignments), and location is of less importance. This duality arises because the British Egg Marketing Board, established in 1957, must purchase all eggs offered to it, but unlike the milk marketing boards, cannot compel producers to sell to it. Since consumers will pay more for farm eggs, which they believe to be fresher or produced by hens kept on free range, some 40 per cent. of eggs by-pass the Board and are sold mainly at farms. There are thought to be about a quarter of a million producer-retailers, and such direct sales are commonest around towns and cities where consumers have easy access to producers, especially in the Midlands and in north-east England, where over 50 per cent. of eggs are sold direct (though the proportion around London was only 15 per cent., presumably because of the size and complexity of its urban area and the importance of large-scale producers).

The remaining 60 per cent. are bought by the Board through 326 packing stations (1967), 267 in England, 34 in Scotland and 25 in Wales; their distribution is mapped in the report of the Re-organisation Commission for Eggs. They are widely distributed and many in Scotland and Wales are small; the average throughput of English stations was nearly $2\frac{1}{2}$ times larger than that of Welsh stations in 1967–8. Packers collect from producers and receive a standard payment for handling and transport costs within 30 miles of the packing station. They decide daily what proportion of the eggs they want and the Board then reconsigns the remainder to other packing stations or, if they cannot be sold, to cold store or to one of the 18 processing plants; the Board pays the cost of this transport. Packers get a similar allowance towards the cost of delivery

K

within 30 miles of the packing stations and the actual cost of transport for greater distances. In effect, the costs of transport are pooled: since each producer receives the same price for his eggs irrespective of his location, those nearer the major urban markets are subsidising the transport costs of more distant producers. Conversely, a buyer in London obtains eggs at the same price from Cornwall or Cumberland as from the Home Counties. On the other hand, packers are not obliged to collect eggs and some have decided not to do so from small or remote farmers. The Board is now to be abolished, and this will probably lead to a greater concentration of production near the major urban markets.

CONCLUSION

Although this brief survey has not revealed much specific information about the patterns of movement of agricultural commodities from farm to market, or about the effects of differences in costs on the choice of markets, it has indicated the complexity of market relationships and the difficulty of making valid judgments about the significance of transport costs in particular areas. Even when such costs are ascertained in different areas and for different commodities, it is difficult to determine whether any apparent modifications of rational transport charges are harmful. It is alleged, for example, that present marketing arrangements often involve unnecessary costs and journeys, e.g. canalising horticultural produce through Covent Garden and fatstock through auction markets, or collecting milk from farms on the upland margins; but always the issues are complex. There are obvious advantages in assembling the wide range of commodities Covent Garden has to offer and, given the varied nature of the produce and of transport charges, according to load, distance and individual, it is difficult to determine whether additional costs have been borne. In livestock marketing, note must be taken of the savings in joint costs with store markets and of farmers' mistrust of sales in which they do not personally participate, although participation is largely necessitated by the poor development of sale by description. In dairying, the limited choice of enterprises on the upland margins, the natural advantages of western districts for milk production and the economics of rationalisation of milk collection must also be considered.

Evidence about the level of transport costs is scanty. In the United States, where transport of agricultural produce from farm to market may involve long distances, transport costs were estimated at 7–8 per cent of farm sales in 1939. No comparable figures exist for Great Britain, but evidence to the Verdon-Smith Committee suggests that such costs might represent 1–2 per cent. of the price received by farmers at fatstock auctions, i.e., the point of first sale. For cereals, differences in transport costs may be considerable; for Scottish barley used in Scottish distilleries it can be as low as 10s. a ton, compared with 50s. a ton, or 10 per cent. of the guaranteed price, for English barleys. The Milk Marketing Board calculated the cost of transporting milk in England and Wales in 1966–7 at 1·07d. a gallon sold wholesale to the Board, or 4·5 per cent. of the net price paid to all producers, ranging from 6·5 per cent. in South Wales to 3·0 per cent. in south-east England. An estimate for 1939, before rationalisation of milk collection, gave 9 per cent. Figures collected by the Runciman Committee on wholesalers' costs suggested that for traders in horticultural produce who dealt exclusively or mainly by outright purchase or sale, transport costs represented 1–4 per cent. of sales according to size of business. The Committee also noted that these costs appeared to be appreciably higher than in most distributive trades, presumably because the produce cannot easily be stored or assembled in bulk. Such figures suggest that while transport charges are not a major cost, they are not negligible; if they can be reduced without loss of other advantages, producers will benefit if this is done. Moreover, these represent averages and there may well be particular commodities, areas or kinds of production where transport costs represent a considerably greater proportion of expenses. For example, the cost of transporting new potatoes by rail from west Cornwall was estimated at 19 per cent. of the selling price in the mid-1950s and that of transporting store sheep from Shetland was 11 per cent. of selling price, compared with only 1 per cent. in non-crofting counties in Scotland and only 2 per cent. in crofting counties as a whole. Similarly, M. Chisholm has shown that differences in the cost of transporting milk to depots, due to variations in the volume of milk collected per unit area, may represent up to 10 per cent. of the price the Milk Marketing Board will realise on marginal output which must be sold for manufacture.

Improvements in transport and the increasing volume of other inputs have certainly lessened the importance of location in relation to markets, while the growing complexities of marketing arrangements and pricing schedules make it increasingly difficult to identify costs. Pricing policies for inputs have similarly weakened the pull of market location; for, although the urban centres are both the markets for farm produce and the source of many non-farm inputs, such as machinery, the price of the latter is either uniform throughout the country or varies less than do the costs of transporting them to farms. Although it is estimated that only 5 per cent. of purchases of farm produce are by direct sale, contract sales to commercial organisations such as Marks & Spencers, which bought £45,000,000 worth of produce in 1963–4, and the growing importance of producer groups and co-operatives, such as the West Cumberland and Yorkshire group, founded in 1964 and possessing 33,000 members in 24 counties in 1967, will complicate marketing relationships still further in the near future.

FURTHER READING

G. R. Allan, *Agricultural Marketing Policies* (1959)

D. K. Britton, *Cereals in the United Kingdom* (1969)

R. Gasson, *The Influence of Urbanisation on Farm Ownership and Practice*, Studies in Rural Land Use No. 7, Wye College (1966)

J. H. Kirk, 'The structure of agricultural marketing in the United Kingdom', *Br. J. Marketing* 1 (1968)

L. V. McEwan, *The Marketing of Store Livestock in Scotland* (1962)

T. K. Warley, *A Report on the Marketing of Milk, Fatstock, Eggs, Potatoes, Tomatoes, Apples and Pears in the United Kingdom* (1962)

Report on Fatstock and Carcase Meat Marketing and Distribution Cmd. 2282 (1964) (Verdon-Smith Report)

Report on Horticultural Marketing Cmd. 61 (1957) (Runciman Report), *Report of the Reorganisation Committee for Eggs*, Cmd. 3669 (1968)

CHAPTER 7

Land and Livestock

This chapter reviews briefly the main categories of agricultural land and livestock, in both their contemporary and their recent historical settings, to provide a perspective for the discussion in Chapters 8 to 13 of the principal enterprises and a link with the preceding chapters in which the factors of production were considered individually. Its brevity does not measure the importance of these topics; it simply reflects a decision that those aspects of the agricultural geography of Great Britain which have already been treated at length elsewhere should receive less attention.

The area of agricultural land in Great Britain measures nearly 46 million acres; a quarter is devoted to crops other than grass and the remainder is almost equally divided between grassland and rough grazing. Each of these land uses predominates in a different part of the country, with most of the rough grazing in Wales, northern England and Scotland, most of the grassland in the western lowlands and most of the tillage in eastern counties. Although there have been considerable fluctuations in their extent, both in recent times and in the more distant past, the main features have been fairly constant over the past two centuries and broadly reflect the regional contrasts in relief, soils and climate discussed in Chapter 3. They are also reinforced by the differences in farm and field size which are the subject of Chapter 4.

TILLAGE

Using land for annual crops is more risky and requires a greater expenditure per acre on field operations than keeping it in grass, although it produces more calories of human food; in theory, an acre of potatoes could support the equivalent of

3·45 persons and an acre of wheat 2·7, compared with 0·45 for an acre of grass used for milk production and 0·16 used for beef cattle, although in production of dry matter or starch equivalent per acre, well-fertilised grass is more productive. The variability of weather is more critical than in grass farming, both because of the need for favourable conditions during the preparation of the land and at the time of planting and because the weather at harvest is more critical for return per acre than on grassland, where hay, when made, represents only part of the yield from grass, and livestock can be protected from the weather or given supplementary feed. In general, the two principal considerations are whether land can easily be cultivated and whether an acceptable crop can be harvested in sufficient quantity and with sufficient regularity to justify the risk, although the individual farmer has to consider many other factors, including the layout of his farm, the equipment and labour at his disposal, and the likely return from different enterprises. Gently undulating land appropriate to the use of large machines and soils that are workable but not excessively drained represent the ideal, and conditions approximating to these are fortunately widespread in those eastern counties where climatic conditions are most favourable for crop production, even if, as in the Fenland and the Humber warplands, these characteristics have been created by human intervention. Another man-made advantage is that the proportion of large farms and large fields is higher than in western counties, and current trends are accentuating these regional differences (Chapter 5); the availability of specialised outlets, such as canneries, and investment in specialised machinery, such as potato and sugar beet harvesters, help to maintain these advantages for crop production. Before the modern period, when the element of subsistence in farming was greater, cropped land was more widespread, but there has long been a contrast, noted by J. Caird in his 1851 survey, between the grazing counties of the west and the corn-growing counties of the east, and this difference has become greater in the past decade with an increasing proportion of the acreage under tillage located in eastern counties. Figure 9, recording the proportion of agricultural land under tillage, shows the importance of that part of eastern England extending from the Yorkshire Wolds to Kent,

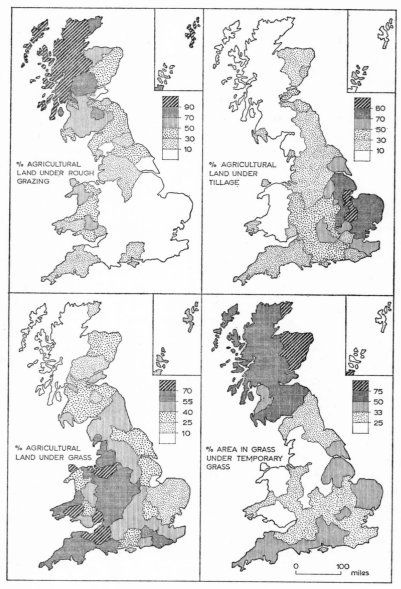

Figure 9. Agricultural land use
Source: agricultural censuses 1965

with an outlier on the chalk downs of southern England; the map of the proportion of improved land under tillage is very similar, although omission of the large areas of rough grazings exaggerates the apparent importance of tillage crops in western districts. As Figure 20 indicates, a map of cash crops would emphasise the importance of eastern districts even more strongly. Of course, such statistical maps only crudely represent the extent of land under crops, for the principal areas contain many pockets of land difficult to plough because of a high water table or impeded drainage, or because soils or relief are unfavourable; but probably the conditions that have encouraged this increasing specialisation in crop production in eastern counties have ensured that virtually all the ploughable agricultural land in these areas is now under tillage, mainly for cash crops. Many areas in these counties, formerly thought to be marginal for arable cultivation because their soils were too heavy or too light and often semi-derelict or abandoned in the 1930s, are now almost entirely under the plough, as parts of the chalklands and the boulder-clay plateau of west Cambridgeshire and Huntingdonshire demonstrate; for the heavy applications of fertilisers now universal on tillage crops and the much greater tractive power available make cropping on both light and heavy soils more feasible. Indeed, the thin chalk soils of the downs of southern England are now highly regarded for mechanised cereal-growing.

The role of soil and relief becomes more restrictive as annual rainfall increases, and heavy land in particular becomes more difficult and risky to work; the main arable areas in western counties are on light soils. Crop production is increasingly concentrated in areas of low rainfall and relatively few rain days, mainly east of the 30" isohyet. This is not to say that climatic conditions are ideal, and irrigation is increasingly used to ensuring that sufficient water is available for high-value crops. The British climate is too variable to guarantee good weather at harvest and thunderstorms are not uncommon in southern England in late summer; cereal harvests can also be seriously handicapped by long wet periods in August and September, and the late onset of growth and the lower summer temperatures in the north-east may delay harvests and create conflicts with autumn ploughing. Nevertheless, within Great

Britain the areas most favourable to cash crops lie in the eastern counties.

GRASSLAND

The second major category of agricultural land is grassland, and grass is the most important of all crops in Great Britain and the one best suited to large tracts of the lowlands. Conventionally, the grass crop is divided into permanent grass, which in theory is never ploughed, and temporary grass, which is part of the arable land and occupies a regular place in rotations. This distinction is still incorporated in farm leases, and nineteenth century leases often specified heavy penalties for ploughing permanent grassland. These categories of grassland have always been less appropriate in western and northern counties, especially in Scotland, with its long tradition of ley farming, in which land is sown to grass for several years after three to five years of tillage crops and nearly all improved land is technically arable. Between 1939 and 1967 the government provided incentives for the wider adoption of ley farming by grants for the ploughing of older grassland (although only in the 1950s did the encouragement of ley farming become a specific aim), and the practice is now thought to be sufficiently well established to justify ending these grants. Advances in grassland husbandry have also encouraged a more flexible attitude towards grassland, so that the distinction between permanent and temporary is increasingly difficult to define. Indeed, the inclusion of questions in the Scottish agricultural census for 1959 about the age of grassland led to a 10 per cent. increase in the acreage returned as temporary, presumably because farmers had not previously considered that older grass should be returned as temporary. The terms permanent and temporary were therefore dropped in the following census and replaced by two categories of grass, under seven years old and seven years old and over. They have been retained in England and Wales, but even here the distinction is somewhat artificial, depending as it must on farmers' intentions rather than on objective external criteria. Undoubtedly a good deal of grassland is never or only rarely ploughed either because it is unploughable or because it is difficult to cultivate; but much land now regarded as permanent grassland has been arable at some time, as the evidence of ridge and furrow on Midland pastures and the records of the ploughing campaign during the second World War show. We shall, therefore, first consider grassland as a whole.

Grassland is most abundant in the western lowlands, notably the Cheshire Plain, the Somerset Levels and south-west Wales, which have a well-distributed rainfall of between 30″ and 40″ and hence are well-suited climatically to the growth of grass. In the drier areas of eastern Britain, where the emphasis is on tillage crops, there is a greater risk that a deficit of moisture in summer may lead to a serious check in grass growth, particularly on the lighter soils; conversely, in those western and northern areas where rainfall exceeds 45″ and soils are generally acid, the farmer often has little choice and the quality of grass is poor. Within these broad climatic limits, the local distribution of grassland is related to the character of the soil and the relief; most of the grassland in eastern counties is located in areas with a high water table, as in the Lincolnshire marshes, or intractable soils, as on the heavy land of the Low Weald, although good drainage is desirable for grass of high quality.

The main use of grassland is to provide feed for cattle and sheep, whether as grazing or as conserved grass, and grass supplies an estimated 70 per cent. of their feed requirements; pigs and poultry eat little or no grass. Among the grazing livestock, sheep are the most dependent on grass, followed by beef cattle; dairy cattle consume more purchased concentrates, although good grass management could greatly reduce this dependence. The value of grazing thus depends on both natural conditions and management. Since grass, unlike most tillage crops, occupies the ground throughout the year, the length of the growing season has some importance and one objective of management is to extend it. As shown in Chapter 3, the season of grass growth is longest in the south-west and decreases northwards and with altitude; the onset of growing conditions also becomes progressively later. Locally the grazing season is affected by the nature of the soil; the onset of growth is delayed on cold, wet soils, and the risk of poaching, i.e. damage by treading (especially by cattle), further limits the effective grazing season on heavy land, although grazing by sheep is generally possible.

About a third of the grassland is also mown to provide feed in winter when growth has ceased and cattle are generally excluded from the fields; the proportion is highest in the south, notably in counties such as Dorset, Somerset and Wiltshire, and declines northward and with altitude. Although climatic conditions are often not very favourable for hay making, about four fifths of the mown grass is made into hay. Silage making is less affected by adverse weather and can use the early grass, which is highest in protein; but although it has been encouraged by a government grant towards the cost of building silos and has increased greatly since the 1930s, it is still a minority interest.

The distribution of permanent grass is similar, although proportions in eastern counties are lower. Apart from the physical controls already noted, farm layout is also a factor, for there is evidence that the proportion of land under permanent grass is related to field and farm size, the former because of the difficulty and expense of cultivating small fields and the latter because of the economies of scale in crop production and because many occupiers of small farms are part-time farmers. In detail, fields may be kept in grass to permit access to other parts of the farm or because they are remote from the farmstead, and machinery cannot easily be taken to them.

The quality of permanent grass varies greatly and improving it offers one of the best opportunities for increasing the productivity of British agriculture. At one extreme are the high-quality fatting pastures of the Welland valley, which are among the most skilfully managed of all pastures; at the other, poor grassland on the upland margins, with a low proportion of nutritious species, is not easily distinguishable from rough grazing. No comprehensive re-assessment has been made since Sir George Stapledon and Dr. William Davies undertook their pioneer reconnaissance survey in 1939, but their results probably still indicate the relative proportions of the different grades and their distributions, despite the wartime ploughing of much of the poorer grass. Stapledon and Davies, using the proportions of perennial ryegrass and common bent in the fields they sampled as indices of good and poor quality grassland respectively, estimated that only 7·4 per cent. of the 15 million acres of permanent grassland in 1939 were in the two highest grades of ryegrass pastures, and that only 35 per cent. had more than 5 per cent. ryegrass. Their map, showing areas where grassland of different qualities is likely to be found, was published in the Ordnance Survey's ten-mile series and shows that, if those areas where permanent grass is unimportant are excluded, the best pastures were in Cheshire, Somerset and the East Midlands, with outliers in Romney Marsh, Holderness and north Northumberland. The poorest quality grassland, dominated by common bent, characterised the acid soils of the south-west peninsula, Wales and the uplands of northern England. Undoubtedly, despite the greatly increased application of lime and fertilisers, there is still much poor grassland, and a post-war survey classified over half the grassland in England as inferior.

Temporary grass is more uniform in quality than permanent grass, for it is generally ploughed when the sward deteriorates; it is also more productive, with higher yields of hay and higher carrying capacity than most permanent pasture, although its establishment involves both expenditure and risk, particularly in a dry season, and temporary grass is more likely than permanent pasture to be damaged by treading. But the relative merits of temporary and permanent grass cannot be judged purely by yield; not only does grass reduce the danger of pathogens accumulating in the soil, but it also improves soils structure and increases yields from the succeeding crops.

The proportion of temporary grass varies widely throughout the country; it is highest in Scotland, where it accounts for two thirds of the grass acreage, and in eastern and southern England. Both the average life of leys and their ratio to the acreage of tillage crops also vary. In eastern England much of the acreage under temporary grass is in one-year leys of annual ryegrass and red clover, but the proportion of longer leys increases westward; in Scotland, too, the grass break ranges from some two–three years in eastern counties to as much as ten further west. The proportion of arable under leys also tends to increase from east to west, where they account for more than half the arable (Figure 9). Largely for these reasons, no very clear pattern of distribution emerges.

Although temporary grass is generally incorporated into some system of alternate husbandry, current trends in arable farming are reducing flexibility, for the removal of hedgerows and the enlargement of fields to permit more efficient crop production make it increasingly difficult to keep grazing livestock on crop farms. Grass, however, provides a desirable break crop in some areas of large-scale cereal growing where alternative cash crops are lacking, and has favoured the combination of dairying and large-scale barley growing in parts of the Hampshire and Wiltshire chalklands.

ROUGH GRAZING

The 17 million acres of rough grazing in Great Britain are largely confined to the hills and mountains of the north-west, for the lowland heaths have been either reclaimed or withdrawn from agricultural use to provide space for outdoor recreation. Although evidence of past ploughing is surprisingly widespread, by far the greater part of the land under rough grazing has never been under cultivation, and most of its fluctuations in area have been along the moorland edge. Nevertheless, large tracts have been improved by cutting open drains and, to a lesser extent, by fertilising and reseeding, and much is regularly burnt to destroy old vegetation and encourage new growth. The higher land is only lightly stocked with sheep, grazing pressures are rather uneven, and little of the land above 3,000' is of any agricultural significance. Except in the New Forest, the rough grazings in the lowlands normally occupy small acreages, generally on steep or low-lying land, and may differ little from poorer permanent pasture in western districts.

Rough grazings are most extensive in Scotland, where they occupy two-thirds of the land used for agriculture, mainly in the Highlands, which contain the most extensive areas of moorland both over 2,000' and under 500'; for in north-west Scotland, moorland reaches sea level and there are only small patches of cultivated land around the coast. The Southern Uplands, the northern Pennines and Wales also contain large blocks of rough grazing and there are smaller tracts in the south-west peninsula. Most of these rough grazings are unfenced or crossed only by boundary walls, although Wales has a large area of enclosed rough grazing, or *fridd*, which is intermediate between the improved land and the open moorland.

The term 'rough grazing' includes a wider range of vegetation and of land qualities than either grassland or tillage, from poor upland grazing that is only nominally agricultural to grassland not easily distinguished from permanent pasture. Generally, this is land which suffers from some severe climatic or physical handicap and has not been thought worth improving in existing technological and economic conditions, although large parts are thought to be improvable with known techniques. The types of vegetation and their characteristics are briefly described in Chapter 10, and reconnaissance maps showing their distribution have been published in the Ordnance Survey's 10-mile series. The vegetation is generally of low nutritional value and makes abundant growth only during the short summer, when it may exceed the ability of the grazing livestock to keep pace; in winter it provides merely enough for survival and winter feed is a major problem for all upland farms. Before the great extension of sheep farming in the last two hundred years, it was solved by the practice of transhumance, which linked lowland pasture and summer grazing by seasonal movements of stock; but the conversion of common grazing into new hill farms with little improved land has separated upland grazings from their complementary lowland pastures, and such movements survive only in the practice of wintering ewe hoggs away, which is declining because of its cost and the increase in winter feeding on hill farms.

The physical handicaps of the upland rough grazings are often aggravated by the structure of farming and by the existence of other land uses. A sixth of the uplands are common pastures or common rough grazings, with consequent difficulties of divided control, and even extensive farms are often too small to provide an adequate level of living. The uplands are also areas of sparse population and long-continued depopulation, and the age structure of the population is often ill-balanced, social services poor and labour difficult to obtain. Agriculture also has to share the upland grazings with other uses, notably the keeping of deer and grouse; the latter, in particular, require land to be managed, and the needs of farmer and keeper may clash. Military training and water gathering also have prior claims on agricultural land, and farming may be further handicapped by restrictions on grazing.

The contribution of these three categories of land to agricultural output cannot easily be assessed, for the uplands mainly provide store and breeding stock, which do not enter into the accounts of the 'national farm', and lowland stock depend heavily on purchased feed. What can be categorically said is that the rough grazings provide only a small part, perhaps between 5 and 10 per cent., although it has been argued that the livestock they supply to lowland farms could not easily be replaced.

CHANGES IN AGRICULTURAL LAND USE

The distribution of rough grazing, grassland and tillage has remained broadly similar since the period of high farming of the mid-nineteenth century, although there have been considerable interchanges between them, especially grassland and tillage. On balance, the extent of cultivated land in Great Britain was probably greater in the 1850s than ever before or since. Reclamation of moorland and woodland was still progressing and the extent of tillage was probably at its peak, although there was undoubtedly land, especially on the clay lowlands, which had been cropped in the past but was no longer under the plough in the 1850s. From the 1870s the the events described in Chapter 2 produced a progressive diminution in area of both improved land and tillage, for improved land was being lost to urban development and to rough grazing as a result of the deterioration of hill land. The decline in the tillage acreage was most marked in the English Midlands, although tracts of heavy land in eastern England, as in south Essex and the Weald, were similarly affected. Very light land, on the other hand, often reverted directly to rough grazing. Least inter-change between tillage and grassland occurred where most of the land was already in grass or where, as in eastern England and Scotland, the balance of comparative advantage still lay with arable cultivation. A reversal of trends during the First World War did not persist after the end of hostilities and by 1938 the area under tillage had fallen from over 14 million acres to just over 8, and was probably more localised than ever before.

The Second World War had a profound effect. There was a rapid extension of the area under crops, not only in those counties where there had been the greatest abandonment of arable farming since 1870, but also in western counties where there had been only a small tillage acreage during the period of high farming. At the peak of the ploughing campaign the tillage acreage reached a total only slightly smaller than that recorded in 1870. An unknown acreage of rough grazing was also reclaimed, although it must have been small in relation to the total area of such land.

For a while after the end of the war, the pattern of changes in agricultural land use resembled that after the First World War, with much of the land that had been ploughed in war-time being sown again to permanent grassland, especially in western and central counties; this occurred despite the incentive of ploughing grants and despite the massive increase in mechanisation during the war and after, although the area under crops in the Midlands did remain above pre-war levels. But from the 1950s, divergent trends emerged and, while the acreage of tillage rose again in the country as a whole and was over a million acres higher in 1966 than in 1958, this trend was at first strongest in those eastern counties where the proportion of tillage was already high, while in western counties the downward trend continued. Consequently the acreage under tillage in eastern England is now larger than ever before, while in western counties it is less, especially in Wales, where the 1965 acreage was under half that in 1875 (Figure 10). In England and Wales, the immediate post-war period also saw a steady increase in the acreage under temporary grass, which rose by 58 per cent. between 1944 and 1961; but here, too, there are regional contrasts, and the need to find more land for cash crops has led to a fall in the acreage of temporary grass in eastern England. This further regionalisation of land use is only one example of the trend towards greater regional specialisation which we shall meet again.

As Chapter 12 will show, the yield of crops has been increasing and the productivity of grassland has also changed; for while the acreage under grass has fallen since 1875, and by

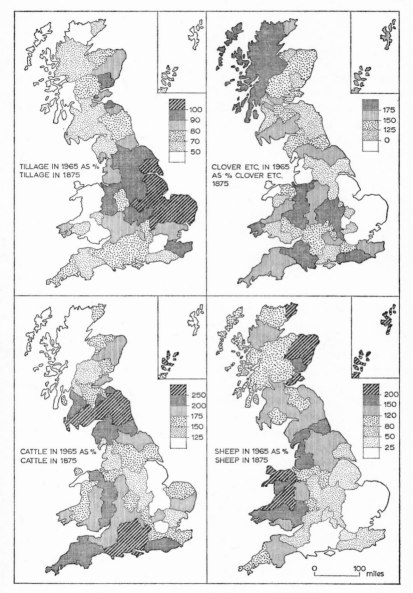

Figure 10. Agricultural changes 1875–1965
Source: agricultural censuses

nearly four million acres since 1938, the numbers of grazing livestock have increased markedly, mainly through a steady rise in those of cattle (Figure 1b), and the density of stocking has increased. This tendency has been marked in the decade since 1955, when the acreage under grass declined to 800,000 acres and the stocking rate is estimated to have risen from 2·20 acres of grassland per livestock unit to 1·87 acres, with the greatest improvement in south-west and south England and the least in north England and Wales. How far these regional improvements arise from better usage of grass one cannot say, for the use of concentrates also increased by some 35 per cent. over the same period; nationally, however, it has been estimated that the productivity of grassland increased by 22 per cent. between 1955 and 1965.

LIVESTOCK

Most books about the geography of agriculture pay particular attention to land use, but in the agriculture of Great Britain this is to consider Hamlet without the Prince of Denmark; for the three-quarters of the agricultural land in rough grazing or grassland provides feed and space for livestock. Further, three-fifths of the acreage of tillage is sown with crops which provide livestock feedingstuffs, whether for consumption on the farm of origin or for sale to other farmers, to merchants or to compounders. Thus, almost 90 per cent. of the acreage of agricultural land is devoted to livestock, directly or indirectly, and livestock and livestock products account for some two-thirds of farm sales in the United Kingdom.

Numbers of livestock cannot be directly compared and must be converted into some common measure, whether as grazing units (cattle, sheep and, where applicable, horses) or livestock units (all forms of livestock). The numbers of livestock units per 100 acres of agricultural land provide a measure of total stocking intensity and show that livestock are most important in absolute terms in the western lowlands, from Lancashire to Dorset, although there are outliers in Anglesey, Cornwall, the Solway lowlands and Romney Marsh; the lowest densities occur, not surprisingly, in eastern England, where cash cropping predominates, and in the northern uplands, especially the Highlands, where both carrying capacity and farm income per acre are low. But when livestock are compared with all other enterprises (Figure 11), the highest proportions of man-days devoted to livestock are found in the western uplands, where 90 per cent. of all labour requirements are attributable to livestock, and the lowest in eastern counties, where crops and horticulture predominate and the proportion attributable to livestock falls below 40 per cent. On this map, too, the simple pattern is disrupted by outliers, as in south-west Lancashire and the south-west Midlands, where crops and horticulture are locally important in predominantly livestock areas.

Cattle are the leading type everywhere except the uplands, where sheep take their place, and the eastern counties, where livestock farming is both more varied and less important and where pigs and poultry account for a larger share of livestock units than elsewhere in Great Britain (Figure 15). If only *grazing* livestock are considered, cattle predominate nearly everywhere except the uplands.

This predominance of cattle has become steadily more marked over the past century, for while the number of sheep has shown a general tendency to decline, superimposed on major falls during the two world wars and subsequent recovery, cattle numbers have risen steadily throughout the period, the increase being particularly steep since about 1930, a reflection of the encouragement of milk production in the 1930s and 1940s and of beef production in the post-war period, although the composition of the national herd has changed considerably. As Chapter 2 has indicated, this rise has depended upon an increase in imported feedingstuffs, especially for dairy cattle, upon the release of land formerly required to support over a million horses, and upon better grassland management. Numbers of cattle rose by 80 per cent. between 1875 and 1965 and, while the number of sheep did not change greatly over this period as a whole, the 1965 total of nearly 29 million represents an increase of 78 per cent. above that in 1947, when flocks, greatly reduced during the Second World War as a matter of policy, suffered heavy losses during the severe winter of 1946–47, to reach their lowest total since regular statistics were first kept (Figure 1). Numbers of pigs and poultry, which can be quickly reduced or increased, have fluctuated considerably, but the

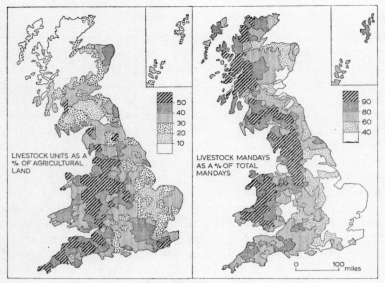

Figure 11. Livestock
Source: agricultural censuses, 1965

Livestock Units. The following factors have been used in computing livestock units: dairy cows, 1; beef cows, 0·8, all other cattle 2 years old and over, 0·75; cattle 1 year old and under 2, 0·5; cattle under 1 year old, 0·25; sows for breeding, 0·5; boars, 0·4; other pigs, 0·25; breeding ewes, 0·2; rams, 0·1; other sheep, 0·66; poultry 6 months old and over, 0·02; poultry under 6 months old, 0·005

Mandays. The following factors have been used in computing mandays: wheat, barley, 2; oats, mixed corn, rye, beans and peas for stockfeeding, mustard, other crops not specified, 3; potatoes, 16; turnips and swedes for stockfeeding, 9; mangolds, 11·5; rape, 0·75; kale for stockfeeding, 2·0; cabbages etc. for stockfeeding, 6; sugar beet, 12·5; hops, 100; orchards, 33; bare fallow, 0·5; lucerne, grass for mowing, 1·0; grass not for mowing, 0·25; rough grazing (sole occupation), 0·1; dairy cows, 12; beef cows, calves under 6 months old, 3; dairy heifers, bulls, 7; beef heifers, all other cattle 1 year old and over, 2·5; calves 6 months old and under 1 year, 2; breeding pigs and boars, 4; barren sows, other pigs 2 months old and over, 0·5; sheep 1 year old and over, 0·75; sheep under 1 year old, 0·25; fowl kept for breeding, 0·17; hens kept for egg-laying, 0·25; broilers, 0·01; other table birds and growing pullets, 0·05; other poultry, 0·1; strawberries, 70; raspberries, 90; red currants, 50; other currants and gooseberries, 45; loganberries, 80; brassicas for human consumption, 25; early carrots, climbing runner beans, 80; main crop carrots, 12; parsnips, turnips and swedes for human consumption, beetroot, lettuce, 35; salad onions, 120; drying onions, 70; broad beans, bush runner beans, green peas for market, other vegetables not specified, 50; other peas, 3; French beans, 130; celery, rhubarb, 40; bulb flowers, 100; other flowers, 250; crops under glass, 1500

general trend has been upwards. Pigs have trebled since 1875 and fowls, first recorded in 1884, have increased more than seven-fold since 1908.

As with the changes in land use on the national farm, the changing composition of the national flocks and herds reflects changes also in their distribution, again towards regional specialisation. Sheep, once widespread throughout the lowlands, have become progressively localised in and around the uplands of western England and Wales, and numbers have fallen most in those counties where folded sheep on arable land were formerly most important, notably in East Anglia and the southern chalk downs of Dorset, Hampshire and Wiltshire. In the Scottish uplands numbers have changed little and have even fallen in places, while sheep have become increasingly important in the counties of north-east Scotland. Cattle, too, have increased more rapidly in the western lowlands, with much smaller increases in the arable counties around the Wash and in the Highlands, although regionalisation has gone much further in the past two decades, with a marked reduction in the number of cattle in most counties in eastern England.

These changes hide changes in the composition of the flocks and herds, particularly in the age of animals; there are also sometimes contrary regional trends. The reasons for these changing patterns are complex and will emerge in part when the individual enterprises are discussed in succeeding chapters. This brief review is a reminder that the situations analysed there in more detail are not static, but merely snapshots of an almost kaleidoscopic pattern of change.

FURTHER READING

J. T. Coppock, 'The changing arable in England and Wales' in R. H. Best and J. T. Coppock *The Changing Use of Land in Britain* (1962)

Milk Marketing Board, Economics Division, *The Balance of Arable and Livestock Farming in British Agriculture* (1967)

CHAPTER 8

Dairy Farming

Milk is the most important product of British agriculture. Sales of milk and the small quantities of dairy products made on farms account for 22 per cent. of farm sales by value and a third of those of livestock and livestock products. Registered milk producers in Great Britain numbered some 95,000 in 1968 and 44 per cent. of all full-time holdings are classified as dairy holdings, nine tenths of them in England and Wales, where 90 per cent. of all milk is produced. Furthermore, from the national dairy herd come most of the cattle slaughtered for beef and veal (Chapter 9).

This dominance has emerged only gradually during the past century, during which the number of dairy cattle more than doubled and production of milk rose more than three-fold; for British farmers have always enjoyed a high degree of natural protection for liquid milk, and dairying was widely adopted from the 1870s onwards by farmers whose location favoured the selling of milk. Dairy farmers, too, who had formerly made butter and cheese, came increasingly to depend on sales of milk to the urban population, and manufacture (no longer on farms) is now primarily a safety mechanism to ensure that enough milk is always available for the liquid milk and fresh cream markets. This dependence on sales of fresh milk is unique among countries where commercial dairying is significant: under the free trade policy the United Kingdom adopted in the nineteenth century butter and cheese could be imported without restriction, whereas liquid milk was protected against competition by its bulk and perishability. The small size of Great Britain and improvements in transport also facilitated production for the liquid market, as did the system of collective marketing adopted in 1933 (Chapter 6); milk can now be carried to urban markets from virtually anywhere in the country and the pricing policies of the marketing boards have further weakened the importance of location in relation to markets for liquid milk.

THE DISTRIBUTION OF DAIRY CATTLE

Dairy cattle are mainly found, as they have long been, in the western lowlands of Great Britain, and the centre of gravity of milk production is moving further westwards (Figures 12 and 20). There are four main areas, each with a tradition of dairying. Easily the most important area, with a fifth of all dairy cattle, lies between the Pennines and the Welsh Uplands, centred on Cheshire, the leading county for dairy cattle; it corresponds broadly to the Milk Marketing Board's North-western Region. Second, with nearly an eighth of all dairy cattle, are the counties of Dorset, Somerset and Wiltshire, comprising the Mid-western Region. The remaining areas are south-west Wales and south-west Scotland, with Carmarthenshire and Ayrshire respectively the leading counties. Dairy cattle are fewest in the eastern and northern counties of Great Britain. The number of holdings with dairy cows reveals a similar pattern, although this information is available from the agricultural census only by regions, which do not correspond with those of the milk marketing boards, and includes a large number of holdings on which only one or two cows are kept to produce milk for the farmer and his staff.

Regional differences in the average size of herds are also marked, and for these the most useful statistics are those of the milk marketing boards, which relate only to farms occupied by registered milk producers. The 1965 Scottish Dairy Census showed that the average herd in Scotland contained 45 cows, with little difference among the three board areas; but Scottish herds were almost twice as large as those in England and Wales, where there was much greater regional variation. Thus, herds in North and South Wales averaged 17 and 18 cows respectively, but those in eastern and southern England were almost as large as Scottish herds and the average for the South-eastern region was 39. These differences are of fairly long standing; although average size of herd rose from 15 cows in 1942 to 26 in 1965 in England and Wales and from 31 to 44 in the Scottish Milk Marketing Board area, the rate of increase

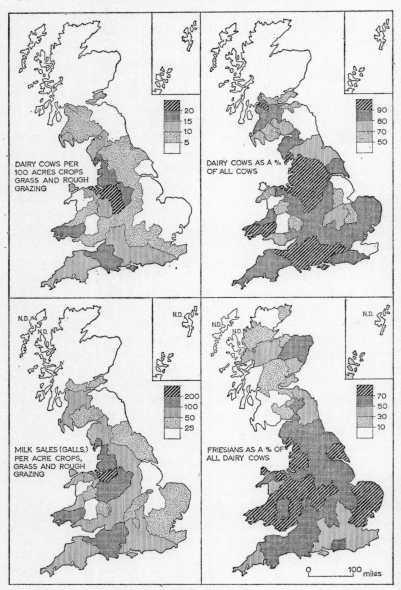

Figure 12. Dairying
Source: milk marketing boards and agricultural censuses 1965. 'N.D.' means 'no data'

was fairly similar in all regions. The principal reason seems to be regional differences in farm size; for most small herds are on

TABLE 11
DAIRY HERDS BY SIZE GROUPS, 1965

Herd Size (Cows)	Northern	North-western	Eastern	East Midlands	West Midlands	North Wales	South Wales
			Average herd size (cows)				
	23	27	30	28	29	17	18
			Percentage of herds				
−10	14	12	18	9	15	31	25
10–19	39	29	29	27	26	42	39
20–39	33	39	28	43	36	21	31
40–59	9	13	12	14	15	4	4
60–	5	7	13	6	9	2	1
ALL	100	100	100	100	100	100	100
			Percentage of cows				
−10	4	3	3	2	3	11	7
10–19	24	15	14	14	13	34	32
20–39	38	40	26	43	35	32	45
40–59	18	21	18	24	24	12	11
60–	16	21	39	17	25	11	5
ALL	100	100	100	100	100	100	100

	Southern	Mid-western	Far Western	South-eastern	Scottish M.M.B.	Aberdeen and district M.M.B.	North of Scotland M.M.B.
			Average herd size (cows)				
	37	36	20	39	44	51	44
			Percentage of herds				
−10	10	11	18	9	3	1	3
10–19	20	17	40	14	9	6	12
20–39	36	38	34	38	36	37	37
40–59	18	19	6	23	30	30	25
60–	16	16	2	17	22	26	23
ALL	100	100	100	100	100	100	100
			Percentage of cows				
						Scotland	
−10	1	2	5	1		1	
10–19	8	7	29	5		4	
20–39	27	30	45	28		27	
40–59	24	26	15	28		31	
60–	40	36	6	38		38	
ALL	100	100	100	100		100	

Source: milk marketing boards.

small holdings and most large herds on large holdings. There is a wide range of sizes in all regions, but Table 11 confirms the evidence of average farm size and shows that most herds in

Wales and, to a lesser extent, south-west England are small and that in herd size eastern and southern England have more in common with Scotland than with Wales and western England. The proportion of herds in different size groups is also changing, as small herds decline in numbers and large herds increase. Between 1955 and 1965 the proportion of herds in England and Wales with under 20 cows fell from 69 to 47 per cent. and their share of cows from 40 to 21 per cent., while those of herds of 50 or more cows rose from 4 to 11 per cent. and 16 to 31 per cent. respectively. Similar changes are recorded in Scotland.

Regional differences in breed structure further qualify the simple distribution map, although these are becoming less marked (Table 12 and Figure 12). At the 1965 dairy census,

TABLE 12
BREEDS OF DAIRY COWS, 1965

Breed	Northern	North-western	Eastern	East Midlands	West Midlands	North Wales	South Wales
			Percentage of dairy cows				
Friesian	62	70	73	68	72	74	80
Ayrshire	22	18	9	16	11	10	5
Channel Island	3	4	9	8	8	3	6
Other	13	8	9	7	9	13	9
ALL	100	100	100	100	100	100	100

Breed	Southern	Mid-western	Far Western	South-eastern	Scottish M.M.B.	Aberdeen and district M.M.B.	North of Scotland M.M.B.
Friesian	50	58	52	50	14	55	29
Ayrshire	19	20	13	19	82	41	62
Channel Island	24	11	19	23	1	1	4
Other	7	12	16	8	3	4	5
ALL	100	100	100	100	100	100	100

Source: milk marketing boards.

Friesians accounted for 64 per cent. of all dairy cows in England and Wales, compared with 41 per cent. in 1955, and were the leading breed in every region; in no region was the percentage less than 50 and in South Wales it reached 80. The Ayrshire, with 16 per cent., was the second most important breed everywhere except South Wales and the Far Western Region, although cows of all Channel Island breeds combined were more numerous in the South-eastern, Southern and Far Western Regions. The proportion of Dairy Shorthorns fell from 25 per cent. in 1955 to only 6 per cent., for this breed has largely been displaced by Friesians. Local breeds such as the Welsh Black and South Devon are still important in their name localities, but less so than formerly. In Scotland, by contrast, the Ayrshire was the most important breed in 1965, with 78 per cent. of all dairy cows, especially in the main dairying counties of south-west Scotland, where the proportion reached 90 per cent or more; but in Scotland, too, Friesians were gaining ground everywhere and had become the leading breed in the Aberdeen and District Board area. The rise of the Friesian among the dairy breeds is due primarily to higher milk yields and to the greater suitability of Friesian calves for beef production; Friesians in recorded herds produce some 10–12 per cent. more milk than Ayrshires and their margin of advantage over cows both of other dairy breeds and of dual-purpose breeds is even greater (although milk from Channel Island breeds, with its considerably higher butter-fat content, commands a premium). This change in breed composition has helped to raise milk yields from 560 gallons per cow in 1938–39 in England and Wales to 815 gallons in 1967–68 and from 573 to 809 gallons in the Scottish Milk Marketing Board area. Regional differences in breed also affect stocking rates; for example, Channel Island cattle, which are important in southern England, require a smaller acreage per cow than Friesians.

Another notable feature of dairy breeding over the past two decades is the rapid increase in artificial insemination, particularly in smaller herds, a development greatly encouraged by the marketing boards. It began in the 1940s and nearly a third of matings in England and Wales were by A.I. in 1950; by 1965, the proportion had risen to two thirds and only 10 per cent. of herds relied entirely on natural mating. In Scotland, possibly because of the difference in herd size, artificial insemination has been less popular, accounting for about half of all matings, but it is increasing. Artificial insemination has also helped to improve milk yields and facilitated the increasing use of calves from dairy herds for beef production, for nearly a third of all inseminations are from bulls of beef breed. It has also permitted the number of bulls to be almost halved over the past two decades, so releasing land for other uses.

Dairy herd replacements also merit consideration. The average number of lactations per dairy cow is between four and five and most dairy farmers aim to replace about a quarter of their milking herd each year. Most replacements are bred on the farm, but small herds are generally more dependent on purchases, particularly in areas such as industrial Yorkshire, where farms are small and there is a high turnover of dairy stock, with dairy replacements bought in from elsewhere and sold after perhaps two lactations. Surveys in 1957–58 showed that 73 per cent. of cows were in self-contained herds, with the highest percentages in south and south-east England and in Scotland, the areas with the largest herds. Dairy followers provide only about a quarter of the gross margin per acre obtained from dairy cows, but the arguments in favour of rearing or buying replacements are complex and the justification for rearing is often only indirectly economic; in an enquiry in 1962 in northern England, where 85 per cent. of herds reared all or nearly all their replacements, the principal reason offered was the risk of importing disease.

Since a dairy farmer aims to achieve a lactation per year from each cow (although the average rate is about 80 per cent.), more calves are born than are required as herd replacements; on average a cow produces three surplus calves, of which two will be male. In some areas surplus female calves are reared and sold as herd replacements, but most calves are either slaughtered for veal shortly after birth or sold to farmers elsewhere to be reared or fattened for beef, the proportion depending largely on breed and on the price of calves and of beef (Chapter 9).

Because most herds provide their own replacements and because the liquid milk market requires a regular supply, seasonal differences in numbers are much less marked in dairy farming than in the rearing or fattening of beef cattle, where the animal itself is sold. Apart from the disposal of surplus calves and culled cows and some inter-regional trade in herd replacements, the distribution of dairy cattle varies little throughout the year. The chief exception is south-west Scotland, where some young dairy stock are sent in summer to graze on non-dairy farms, chiefly from smaller farms, a practice that enables them to carry more dairy cows than they otherwise could.

Production, however, shows some seasonality. Milk is most cheaply produced from grass-fed cows in summer and, although calving takes place throughout the year, there is a spring peak when grazing is most abundant (with a secondary peak in autumn when most heifers and young cows calve). Milk production is therefore higher in the summer six months (53 per cent.), reaching a peak in May, but monthly contrasts are much more marked; production in England and Wales in January 1968 was only 76 per cent. of that of May and the differences were even greater in the Scottish Milk Marketing Board area, where November production was only 63 per cent. of the May peak. Seasonality also has a regional component: areas where the climate favours grass growth have an advantage in summer and the main cropping areas (which have arable by-products) a corresponding advantage in winter when cows depend on preserved and purchased feed; in 1965, the proportion of winter milk was above average in the Southern, Eastern, East Midlands and North-western Regions, while summer milk was more important in Wales and Scotland. Nevertheless, no clear regional pattern emerges and Milk Marketing Board records show a considerable range within each region; for production is more markedly seasonal on smaller holdings, on those occupied by older farmers and on those with only a limited dependence on milk (although these characteristics, too, tend to vary regionally). Seasonality of production was much more marked where most milk was manufactured, especially areas which specialised in cheese making, and production was mainly concentrated in the summer months; J. D. W. McQueen has suggested that the marked seasonality in south-west Scotland is inherited from traditional patterns of calving developed in the hey-day of cheese making. Faced with an increasing surplus of milk over the requirements of the liquid milk market, the Milk Marketing Board is endeavouring to concentrate production in the summer months by reducing the difference in seasonal prices, a policy which hurts farmers in eastern England, who specialise in winter milk production.

Each facet of the distribution of dairying has changed during the evolution of the modern dairy industry, and trends towards larger herds, a higher proportion of Friesian cattle and a rising proportion of production in the western lowlands are continuing features. Nevertheless, there is some stability in the pattern of milk production and the principal factors that appear to influence the distribution of dairying will now be considered.

FACTORS INFLUENCING THE DISTRIBUTION OF DAIRYING

The long-continued prominence of dairying in western lowlands suggests that physical conditions are important, and many features of the evolution of dairy farming over the past century reflect the changing balance between the pull of the market and physical suitability. The dairy cow is more sensitive to climatic conditions than other grazing stock; the optimum temperature for milk production is said to be about 50° F. (10° C.) and output can fall considerably in cold weather. Such hazards can be evaded at a cost by housing, and the 1957 dairy herd survey showed length of winter housing increasing from south to north. Environmental conditions, however, probably make their influence felt mainly through the availability and quality of grazing, although grazing accounts for only 17 per cent. by value of the feed consumed (which in turn represents over half the cost of production) and even in summer large quantities of concentrates are fed to dairy cows. The Meteorological Office has classified counties in England and Wales according to their suitability for milk production, using as an index average annual transpiration, which depends on temperature, sunshine, wind, humidity and rainfall. Three groups of counties are distinguished. Those with an annual average transpiration of more than 15·2″, which include the whole of Wales and the Far Western and North-western Regions, are regarded as well suited to milk production, which is increasing in these areas. Counties with transpiration less than 14·0″ are confined to eastern England from Lincolnshire to Kent and are poorly suited to dairying; the remaining counties, with values between 14·0″ and 15·2″ are moderately suited. Other physical factors, such as soil texture, soil moisture, and elevation, can modify these county assessments locally. For example, dairying is practised along the river valleys of East Anglia, where the water table is high. Conversely, large tracts of counties classified as well suited are upland rough grazings, which exposure, high rainfall and poor vegetation make unsuitable for dairying. Climatic suitability is measured by the length of the grazing season, which ranges from some 31 weeks in southern England to 26 weeks in northern England, although the period varies widely within each region. Such variations are partly due to local differences in the physical environment, especially in soil texture, for the effective grazing season is generally shorter and more variable on heavy soils, which are liable to poach in wet weather; thus, most herds on the chalk downlands of southern England remain outside throughout the year while those on adjacent clays are housed in winter. Nevertheless, dairying would undoubtedly be even more strongly localised in the lowlands of the west and south-west if physical suitability were the only criterion.

Farm size is also a factor, although its influence is mainly indirect. Most dairy farms in England and Wales (61 per cent.) are under 100 acres, although they produce only 38 per cent of output. Small dairy farms are especially characteristic of Wales and western England, for dairying is one of the few enterprises capable of providing an acceptable income from a relatively small acreage, particularly where horticulture is not possible; at the beginning of the century, dairying was regarded as particularly suitable for small holdings, and the secure market and regular income provided by the Milk Marketing schemes later made it attractive to occupiers of small farms, especially by comparison with livestock rearing, where there is an uncertain return only at long intervals. Nevertheless, as improvements in labour productivity increase the size of herd that one man can manage, small farms have been placed at a disadvantage. Costs are generally higher for small herds, for they are more dependent on purchased feedingstuffs (which in effect increase their acreage) and on purchased herd replacements, and yields per cow are also lower on average; in the National Milk Investigations the acreage of grass and forage crops per cow in 1965 on holdings of under 25 acres was only 84 per cent. of that on holdings of 100–149 acres. At the same time, the occupiers of small farms are generally more dependent on milk and the smallest holdings are often entirely devoted to milk production. Land is usually the limiting factor on such farms: many of them are simply too small to make enlargement of the dairy herd feasible. Consequently occupiers of small farms have tended to give up dairying in recent years, and a sample survey in England and Wales showed that 66 per cent. of those abandoning dairying between 1963 and 1967 had holdings of under 100 acres, a proportion that would undoubtedly be higher but for the lack of suitable alternatives in many areas.

Such small holdings are both numerically and proportionally less important in Scotland, accounting for only 38 per cent. of all dairy farms in 1965. Scottish dairy farms are large, both absolutely and by comparison with other types of farm. This is partly due to a larger acreage of rough grazing on many Scottish farms, but it is also reflected in the higher average size of dairy herds. Here, too, proportionally more small dairy farms have tended to go out of production.

The labour requirements are both greater and more regular than for other grazing livestock, and the availability of labour increasingly imposes a restraint on keeping dairy cattle, especially in those areas least dependent on dairying; difficulties in recruiting and retaining hired labour are frequently mentioned by farmers as a reason for giving up dairying. It is true that most labour on dairy farms in England and Wales is family labour, especially in western counties: family labour accounted for 86 per cent. of all full-time labour on dairy farms in South Wales in 1963, though only 27 per cent. in the Eastern Region and 62 per cent. in England and Wales as a whole. Yet, even for family labour, the demanding nature of dairy farming provides farmers with a motive for ceasing production, although the lack of suitable alternatives prevents many from doing so. Even some occupiers of larger farms in a survey of those who had abandoned dairying in northern England in 1958–62 were willing to accept lower returns from other enterprises to escape from 'the drudgery of milk production' with its seven-day week.

At the same time, economies of scale in labour use provide an additional incentive to those with small herds either to abandon dairying or to increase the size of their herds; indeed, lower labour costs are the chief element in cost reduction with increasing size of herd. In 1965–66 labour hours per cow in England and Wales ranged from 62 hours in herds of 60 or more cows to 162 in herds with less than 10 cows, compared with an average for all herds of 88; moreover, while this is not new (the averages in 1950–51 being 113, 212 and 142 hours respectively), the reduction has been proportionally greater in the largest herds. Similar differences are also expressed regionally in numbers of cows milked per worker, ranging in 1963 from 11 in South Wales to 27 in the Southern Region; a contributory factor may be the survival of some hand milking in western counties. The occupiers of small farms can neither have a sufficiently large herd to enjoy such economies, nor make more effective use of labour saved by improved labour productivity.

The nature of farm buildings is another factor; for dairying is unusual in that farmers wishing to sell milk must first have their premises inspected and licensed. Lack of suitable buildings is often mentioned as a reason for giving up dairying and may also be important in the decisions of farmers who do not mention it; in one survey in 1964 buildings were given as the reason for abandoning dairying by only 8 per cent. of farmers, but buildings on 20 per cent. of the farms surveyed were either too small or in poor condition, and those on 26 per cent. of the farms were not conveniently situated in relation to the fields used by the dairy herd. Most cows are milked in cowsheds, many of them adapted from buildings designed for other purposes; with bigger herds, economy of labour use increasingly depends on efficient layout, and 58 per cent. of the cows in herds of 70 or more in 1963 were milked in parlours, compared with only 28 per cent. in all herds in England and Wales. Not surprisingly in view of these regional differences in herd size, the proportion milked in parlours was highest in southern and eastern England, averaging 45 per cent. of cows in all herds in the South-east Region, compared with 6 per cent. in South Wales. Only in the Mid-western region are bails important.

The influence of markets on the location of dairying is complex. In early location theory, milk for sale was produced near towns because of its perishability and bulk, and dairy farmers in distant areas concentrated on production of butter and cheese, which were less perishable and less bulky. As Von

Thünen recognised, such simple relationships must be modified to take account of areal differences in suitability for milk production, which are reflected in differences in costs. Further complications arise from the need for a reserve of milk to ensure that supplies for the liquid market are always adequate; from improvements in the treatment and transport of milk, which have extended the area within which it is technically and economically feasible to supply milk to urban markets; and from the organisation of both buyers and sellers of milk, for the movement of milk from farms in Great Britain is now controlled by the four milk marketing boards and its purchase is largely in the hands of a few big companies.

MILK PRODUCTION & MARKETS

The changing relationship between milk production and markets merits an historical digression, which will also place the dairying industry in perspective. For simplicity, and because of the importance of dairying in England and Wales, this discussion will concentrate on those countries.

In the early part of the nineteenth century the pattern of dairying in England and Wales reflected clearly the opposing pulls of market location and physical suitability, although until the second half of the century milk for the urban population represented only a small part of total output of dairy produce. Cows were kept either in urban cowsheds or on farms within a few miles of the town. In 1865, milk cost 6d. a gallon to transport 18 miles (more than the cost for several hundred miles today) and 72 per cent. of London's milk in 1861 came from stall-fed cattle in London itself. Most dairy cattle were in the main areas of grassland farming, especially the western lowlands, where they supplied milk for making farmhouse butter and cheese and for livestock rearing; for most cattle were dual-purpose shorthorns, and the distinction between dairy and beef cows was less clear cut than it is today. This balance between supplies of milk for liquid consumption and for manufacture altered as the urban population increased; as improvements first in rail and then in road transport and also in milk treatment permitted it to be transported safely over long distances and at less cost; and as depression in arable farming and growing imports of dairy produce encouraged farmers to supply milk for the liquid market wherever they could. The cattle plague of 1865 and the decline in prices for fat cows (the source of a substantial part of income from urban dairy herds) were also factors. The effects of these developments are best exemplified by the changing sources of London's supply, although London, because of its size, higher per capita consumption and remoteness from those areas where natural conditions are most suitable for milk production, was (and still is) a special case, and the change here was accelerated by the cattle plague, which reduced local milk supplies by nearly 60 per cent. Nevertheless, the transformation is striking. In 1861 only 4 per cent. of London's milk came by rail and the maximum radius (which related to only a small gallonage) was some 50–60 miles; by 1891, 83 per cent. was carried by rail, total supplies had doubled and the radius of supply had trebled to embrace traditional dairying counties such as Wiltshire and Staffordshire.

This expansion of production for the liquid milk market was achieved partly by diverting supplies from the manufacture of farm butter and cheese, as in Buckinghamshire and Wiltshire, and partly by the adoption of dairy farming in areas where it had not previously been practised, especially where much land, with good access by rail, was being converted to permanent grass. Thus, the distinction between manufacturing areas and those supplying the liquid market was already breaking down and this tendency was accentuated as the need to organise supplies for large urban markets led to the emergence of wholesaling, first in London and then more generally, and to the erection of country depots where milk was bulked and treated for onward transmission and milk not required for the liquid market was manufactured, although much farmhouse manufacture persisted until the 1930s.

For a while the location of urban markets continued to be of some importance. On the one hand, there were still concentrations of dairy cattle in and around the larger towns and, on the other, dairying areas, such as south-west England, were still too distant to supply the liquid milk market; in 1906–14 transport costs for farms more than 100 miles from their markets represented 27 per cent. of the price received by farmers. In 1918 a quarter of all supplies were still provided by producers close to markets who retailed their own milk, and such producer-retailers were estimated to provide half the supply of provincial towns in 1925, especially in the industrial cities of northern England. Most large cities drew their supplies from within a 50-mile radius and long-distance travel, although greatly reduced in cost since the 1860s, was mainly confined to London milk, which accounted for two thirds of all rail-borne supplies in 1925; Newcastle, drawing supplies from south-west Scotland, was the only other city to depend significantly on distant sources. The increasing demand for milk in the Midlands and south Pennines conurbations, however, led to some contraction of the London milkshed in that area and, since by the 1920s all the milk produced within a 40-mile radius of the capital was required for local consumption, London dairymen had to look further west.

Nevertheless, although much milk was produced and consumed locally, London tended to set the level of prices elsewhere; for milk could be diverted there from most of the English lowlands if local prices fell below the London price less transport costs. As transport improved and cheapened, producers near urban markets, whose higher costs for level production throughout the year had been compensated by higher prices, increasingly faced competition from low-cost producers in the western lowlands who had previously supplied the less profitable manufacturing market. For a while in the 1920s collective bargaining between the National Farmers Union and the dairymen maintained prices, but the fall in prices for butter, cheese and other dairy products in the world economic crisis in the early 1930s encouraged further diversion of manufacturing supplies; for example, producers as far away as Wigtownshire were prepared to sell milk on the London market for as little as 4d. a gallon.

In these circumstances the milk marketing boards were established. They were empowered to fix prices on behalf of all producers, which strengthened their position vis-a-vis the large combines increasingly controlling the sale of milk in urban areas, and they adopted marketing schemes which created separate markets for liquid and manufacturing milk and thus removed the possibility of a similar collapse of the liquid milk market. Initially the returns from all milk sold in each of the eleven regions into which England and Wales were divided were pooled to determine an average market realisation or pool price, which was paid to all producers irrespective of their location. The Milk Marketing Board was able to secure a uniform regional price for liquid milk much higher than the prices obtainable for manufacturing milk (which had to compete with imported produce), so that pool prices varied inversely with the proportion sold for manufacturing and were highest in those regions nearest the urban markets. In June 1934, the highest pool price was nearly 60 per cent. above the lowest and such differences would have led to considerable anomalies between farms on either side of a regional boundary; for example, producers in the Southern Region would have received 40 per cent. more than those in the adjacent Mid-Western Region. To prevent undue variation of pool prices between regions, a system of inter-regional compensation provided that regions with above average sales to the liquid market (i.e. those nearer their markets and with no marked summer surplus) subsidised those with below average sales. In 1935 'undue' was interpreted as not more than 1d. per gallon, then representing about 8 per cent. of average pool prices, but in 1942 it was reduced to $\frac{1}{2}$d.; there it has remained, although it now represents little more than 1 per cent. of current pool prices. Location retained some importance in that producers paid transport charges which they negotiated with buyers; those selling to depots or factories paid an additional levy, for they might otherwise have been better placed than those in the same region selling direct to urban markets. Producer-retailers continued to sell direct, but paid a levy to the Board. Some features of the scheme, however, provided an incentive to distributors to seek supplies from more distant producers and a complex pattern of cross haulage of milk developed.

By preventing undercutting by distant producers, the marketing schemes re-established the profitability of dairying in areas near urban markets, although without any premium for proximity; but, more important, by providing the same price throughout a region, and by inter-regional compensation, they increased the attractiveness of milk selling in areas remote from urban markets, and the assured market for milk and regular payment provided an additional incentive to farmers to take up dairying. Following the establishment of the scheme, the number of producers rose steadily and sales of milk off farms increased, especially in Wales and the Northern and Far Western Regions, partly by diverting supplies from livestock feeding and from farm manufacture, which decreased sharply; for example, farmhouse manufacture of cheese fell by two-thirds between 1930 and 1934. Since liquid sales did not increase as rapidly, the proportion of milk sold for manufacturing rose from 25 per cent. in 1933–34 to 31 per cent. in 1938–39.

During the Second World War, modifications of the scheme further reduced the importance of location. Except for the small and declining volume of milk sold by producer-retailers (whose number fell from 29 per cent. of all producers in 1945 to 9 per cent. in 1965, when they sold only 3 per cent. of all milk sold from farms), producers lost any say over the destination of their milk, and the collection of milk from farms was rationalised in 1942, a move which yielded a saving of one fifth in transport. The Board organised the collection of milk from farms and a system of regional transport rates covered the cost of transport from farm to first destination, which might be either a town dairy or a depot. At the same time, the cost of transporting milk from depots to urban markets was borne by the Exchequer.

Regional transport rates varied between regions, but were uniform for all producers in each region, except that between 1943 and 1962, a higher rate ($\frac{1}{2}$d. per gallon until 1947 and $\frac{1}{4}$d. thereafter) was payable by those who sent milk to a depot. These rates were not the actual cost of transporting milk to its first destination, which varied in 1961 from 0·76d. per gallon in the Mid-Western Region to 1·18d. in the Southern Region, for they incorporated a regional differential to give some weight to the proximity of regions to liquid markets; thus, producers in regions near liquid markets paid less than the costs of collection and more distant ones paid more. These differentials, however, are less than the cost of transporting milk from distant regions, which may be as much as $4\frac{1}{2}$d. per gallon; to that extent, the effects of location have been further reduced.

The marketing schemes have thus weakened the influence of urban markets on the location of dairying, although their effects are difficult to assess; for they have been implemented while improvements have been reducing the cost of transporting milk, so lessening the importance of distance, and when collection and distribution of milk have been increasingly rationalised to secure greater efficiency. Although regional transport rates have made it easier for farmers in remote areas to sell milk, the rationalising of ex-farm collection has lessened the cost of transport from farms and made it difficult to determine the true cost of collection from each farm; but the saving from charging actual costs would probably be small. Bulk collection, which minimises handling and so reduces transport costs, and which accounted for 23 per cent. of milk collected in 1967–68, is further reducing the significance of uniform charges within regions. The effect of relieving producers of the cost of long-distance transport from depots, mainly to London, is even more difficult to establish but may represent as much as 10 per cent. of net prices; but it is partly the price for ensuring that milk is always available. It is Board policy to minimise transport costs, and transfers between regions take place only to relieve surpluses or to supply deficiencies. The distribution of manufacturing facilities, which partly reflects the historical evolution of milk production, is also being rationalised to ensure that manufacturing milk does not travel long distances. Recent complaints by producers in south-east England that their natural advantages in proximity are being discounted cannot easily be proved or disproved; but it is certain that they would not receive present prices in free market conditions and that their locational advantage is being eroded by the continued reduction in transport costs.

Of greater significance for the present geography of milk production are the effects on pool prices of the quantity of milk surplus to the requirements of the liquid market. Because milk production is seasonal and supply varies, a surplus of about 25 per cent. above liquid sales over the year as a whole is required to ensure supplies at all times, and this is included in the standard quantity to which the price guarantees apply; any surplus above the standard quantity realises only the manufacturing price and so depresses the price received by all producers. Since 1951, production has risen faster than liquid consumption and an increasing proportion has been manufactured; farmers' costs have risen sharply and, with decreasing profit margins, the number of milk producers has fallen steadily. It is principally those with small herds who have been giving up dairying, especially in eastern counties where there are alternative enterprises. Those who continue are increasing the size of their herds, and numbers of dairy cows have been rising in western and northern England and declining in eastern England; thus, output of milk from Wales and north-west and south-west England rose by 26 per cent. between 1953 and 1963 and the number of cows by 12 per cent., while that of producers fell by 24 per cent.; in the Eastern Region the proportionate changes were respectively an increase of 6 per cent. and decreases of 7 per cent. and 43 per cent. Since any surplus to liquid requirements must be manufactured, this should preferably arise in summer, when costs of production are lower, and in western regions, which are better suited to summer milk production. Some of these changes in production and disposal are summarised in Table 13 which shows the great increase in milk sales in western counties since 1924-5, when regional figures first became available. It also shows the varying dependence of the regions on sales of milk for manufacturing and through depots.

This analysis of the changing role of markets has concentrated on England and Wales, because of the prominence of dairying in these countries. Developments in Scotland have been broadly similar, but the smaller size of the Scottish markets for liquid milk, the greater importance of manufacturing, the unsuitability of much of the country for dairy farming and the location of the major milk market, the Clyde Valley conurbation, in an area favourable for dairying all modify the relationships between markets and dairy farming. In the nineteenth century there were, as in England and Wales, two distinct outlets for milk, liquid milk being produced in and around Edinburgh, Glasgow and other towns, and manufacturing milk mainly in south-west Scotland, where a flourishing farmhouse cheese industry had grown up. From the late nineteenth century farmhouse cheese faced increasing competition from imports and there was a similar tendency for those producing milk for manufacture to invade the liquid milk market, not only in Scotland, but also in London and north-east England. The Scottish Milk Marketing Board (the other boards were

TABLE 13
PRODUCTION AND MARKETING OF MILK, ENGLAND AND WALES

	Northern	North-western	Eastern	East Midlands	West Midlands	North Wales	South Wales	Southern	Mid-western	Far Western	South-eastern
Percentage of total sales in England and Wales, 1924–5	6·8	27·3	6·0	8·4	8·1	2·8	3·5	7·5	14·5	3·7	11·5
Percentage of total sales in England and Wales, 1967–8	10·0	21·4	4·3	5·4	9·7	4·3	7·4	5·8	14·3	10·1	6·9
Percentage increases 1924–5 to 1967–8	389	206	187	171	318	404	566	203	259	721	177
Percentage of milk sold for manufacturing in each Region, 1966–7	37	20	5	11	33	54	39	5	42	39	7
Percentage of milk manufactured in England and Wales, 1966–7	15	13	1	2	15	5	12	1	20	14	2
Percentage of wholesale milk sold direct in each Region, 1966–67	48	76	61	87	47	18	19	91	25	14	100

Source: Milk Marketing Board.

formed later to deal with the special problems of north and north-east Scotland) was created in similar circumstances to the Milk Marketing Board for England and Wales, but instead of regions with uniform prices throughout, three liquid milk markets (Dundee, Edinburgh and Glasgow) were nominated and transport was paid at rates tapering with distance from these centres; later, to meet objections from high-cost producers in eastern Scotland who felt their markets threatened by low-cost milk from south-west Scotland, twelve other centres, mainly small towns in eastern Scotland, were nominated to restore the advantages of proximity.

By agreement between the marketing boards, Scottish dairy farms ceased to supply English markets with milk and south-west Scotland has remained primarily a manufacturing area, supplying milk for the liquid market mainly in winter. Glasgow can still obtain most of its milk from within a 20-mile radius. Because of the shorter distances, there was little need for country depots to bulk and treat milk for onward transmission; most milk now moves direct from farms to urban dairies, while the creameries in the main dairying areas manufacture most of the milk they receive. Bulk collection by tanker, which has gone much further in Scotland (with 79 per cent. of milk in the Scottish Milk Marketing Board area collected by tanker in 1967) has further strengthened this tendency. The location of liquid milk supplies in winter is influenced not only by urban deficits, but also by the kind of manufacturing; thus, milk moved to Edinburgh from Wigtownshire, where cheese is manufactured, and not from Dumfriesshire, where condensed milk (which fetches a higher price) is made, although the latter county is much nearer. Producer-retailers are also more important in Scotland, accounting for 16 per cent. of producers and 6 per cent. of milk in the Scottish Milk Marketing Board area and for 25 and 15 per cent. respectively in the Aberdeen and District Milk Marketing Board area. The greater reliance on the manufacturing market is recognised in the price guarantee, for the standard quantity is some 60 per cent. above the gallonage used for liquid consumption, compared with 25 per cent. in England and Wales. There thus remains a clearer distinction between manufacturing areas and liquid milk areas and a closer link between liquid milk production and urban markets. Nevertheless, similar tendencies are discouraging milk production in areas where there are alternative enterprises. The number of producers has fallen steadily since 1951, and the number of cows since 1955; as in England and Wales, these tendencies are most marked in eastern counties, where some quite large milk producers went out of production in the 1960s.

Thus, only for that small part of the dairying industry which supplied the liquid milk market before 1875 has market location been the primary determinant of the location of dairy farming in Great Britain. As the demand for liquid milk grew, the balance between manufacturing and liquid milk shifted in favour of the latter and market attraction was gradually weakened by technical and economic developments. In these circumstances, the diversion of milk into liquid markets from the manufacture of products, whose price was largely determined by the prices of imports, would have led to the disappearance of many producers located near their markets (in so far as they were high-cost producers); collective marketing prevented this by insulating liquid and manufacturing outlets. The milk marketing boards also reinforced the tendencies that were reducing the relative importance of transport costs, although some residue of locational advantage remains in England and Wales in the differences in regional transport rates. Nevertheless, the greater profitability of dairying in areas of low production costs, rising yields and the technical factors making for herd enlargement have led to an increasing production of milk above the needs of the liquid market, and this, because of the pricing system adopted, depresses the price to all producers. This development has encouraged those producers who have either higher costs or alternative enterprises to abandon dairying; interesting contrasts are shown by East Anglia, where there are profitable arable alternatives and this tendency is most marked, and the Weald, where the only alternatives are extensive livestock enterprises and trends resemble those in the main dairying areas.

TYPE OF FARMING

Not only are dairy farms the chief type of farm (Chapter 14), but dairying is more specialised than other enterprises, in that the great majority of dairy cattle are on dairy farms. In England and Wales, 75 per cent. of all dairy cows were on dairy farms in 1965, and in Scotland 89 per cent. This tendency to specialisation is also becoming more marked.

Dairying has always been more widespread in England and Wales than in Scotland, although all three show an increasing tendency for dairy cattle to be concentrated on fewer farms and in those areas which seem, on physical grounds, best suited to

TABLE 14
DAIRY COWS IN EACH REGION OF ENGLAND AND IN WALES, 1966

Type	England & Wales	Eastern	South-east	East Midland	West Midland	South-west	Yorks & Lancs	Northern	Wales
	Proportion of dairy cows by holding type								
Specialist Dairy	45	17	44	41	54	51	44	27	48
Mainly Dairy	37	33	38	40	34	34	39	49	35
Mixed	12	21	13	12	9	12	10	15	9
Others	6	29	5	7	3	3	7	9	8
ALL	100	100	100	100	100	100	100	100	100
	Average size of herd (cows)								
Specialist Dairy	35	42	43	33	40	36	31	34	28
Mainly Dairy	33	44	48	31	38	35	27	31	24
Mixed	24	35	40	22	27	24	21	20	20
Others	30	35	41	28	36	31	25	26	23
	Proportion of dairy cows in each region								
	100	5	11	8	17	26	10	9	13

Source: Ministry of Agriculture

dairy farming. The proportion of dairy cattle on dairy farms, i.e. those farms on which dairying accounts for more than 50 per cent. of all standard man-days, exceeds 75 per cent. in each of the N.A.A.S. regions except Eastern England and is highest (88 per cent.) in the West Midlands (Table 14). Specialist dairy farms (i.e. those with over 75 per cent. of their standard man-days in dairying) are relatively most important in the West Midlands and the south-west. Except for Eastern England, where 27 per cent. of dairy cows are on cropping farms, most of the remainder are on mixed farms, with proportions ranging from 9 per cent. in the West Midlands to 21 per cent. in Eastern England.

In Scotland, where nearly two-thirds of the dairy cows are in the South-west, the concentration of dairy cows on dairy farms is marked in all regions, with the proportions ranging from 94 per cent. of all dairy cows in the South-west to 73 per cent. in the North-east (Table 15). No other type of farm is important, but the proportion of dairy cows on each of the other types is also highest in the South-west, where the average number on each dairy farm is several times higher than in other regions. The predominance of the South-west is even more striking in respect of other dairy cattle, for average numbers of these are markedly higher on all other types of farms in the South-west than in other regions. This is especially true on hill sheep, upland and lowland farms, where young dairy stock may be sent in summer.

TABLE 15
DAIRY COWS IN EACH REGION OF SCOTLAND, 1962

Type	Scotland	Highland	North-eastern	East central	South-east	South-west
	Proportion of dairy cows on dairy farms					
Dairy	89	86	73	83	84	94
	Average size of herd (cows)					
Dairy	44	32	44	49	39	44
All full-time	13	6	5	8	8	28
	Proportion of dairy cows in each region					
	100	5	13	10	6	64

Source: Department of Agriculture

Herds on farms classified as dairy farms are generally larger than those on other types of farms, with those on mixed farms intermediate in size; thus, while the average for all other full-time farms with dairy cows ranges from 29 cows in South-east England to 14 in Wales, the averages for dairy farms are 45 and 26. There are also considerable regional differences, reflecting the importance of other enterprises; while in eastern and northern England the average herd is larger on mainly dairy farms (i.e. those with between 50 and 75 per cent. of their standard man-days in dairying) than on specialist dairy farms, in western regions the reverse is true (Table 14). Comparable data are not available for Scotland but average herds on dairy farms range from 49 cows in the East Central Region to 32 in the Highlands, compared with an average on all full-time farms of between 5 and 8. In the South-west, partly because of the large proportion of the Scottish dairy herd in that region and partly because dairy cattle are relatively more important on other types of farms, the differences are much less.

Some confusion about the relationship between dairying and type of farming arises from the different use of the term 'dairy farm' by the agricultural departments and by the milk marketing boards, for the latter regard a dairy farm as any farm occupied by a registered milk producer. Thus, while most milk-selling farms are likely to be included in the categories of specialist and mainly dairy farms, on some others dairying is a minor interest; equally, there are farms which have dairy cattle but do not sell milk, and it is estimated that some 6 per cent. of all dairy cows in Scotland are not on milk-selling farms. Nevertheless, in the 1963 sample census conducted in England and Wales by the Milk Marketing Board, dairying was the most important enterprise on milk selling farms in all the Board's regions, with the highest proportion in the main dairying regions and the lowest (62 per cent.) in the Eastern Region; and where dairying was not the leading enterprise, it was nearly always the second. Apart from dairy followers (recognised in this 1963 analysis as a separate enterprise), pigs and poultry were the second enterprise after dairying. Sheep were important as a second enterprise only in Wales, while cash crops provided the leading enterprise on milk-selling farms in Eastern England, and the second enterprise in the Southern and South-eastern Regions. Fattening of beef cattle was nowhere a major enterprise and its importance declined as the size of farm increased. Thus, although dairying is associated very largely with dairy farms, this term includes a considerable variety of types, with farms in eastern and southern England having larger herds and generally larger farms and with cash crops as important subsidiary enterprises, and those in western counties being generally smaller, with more grass and a greater emphasis on dairying. Dairying may indeed be the chief enterprise, but the farm on which it is practised may range from a small, largely grass farm on the lower slopes of the

Pennines, with few, if any, other enterprises, to a large, mainly arable farm on the chalklands of southern England, where a large dairy herd is kept on grass leys and barley provides much of the farm income.

Comparable data are not available for Scotland, but a classification of holdings in 1962 shows a rather similar pattern, with less marked contrasts. Size of both business and average herd is larger on dairy farms in eastern Scotland, although such farms are much less numerous than in south-west Scotland. Crops are likewise more than twice as important on dairy farms in eastern counties as in western, and sheep, generally the second enterprise on dairy farms in Scotland, are most numerous on those in the Highlands and South-east Scotland.

This concentration of dairy cattle on dairy farms, however diverse in character, is long-established and owes something to the earlier association of dairying with small holdings and, since the institution of the milk marketing boards and of price guarantees, to the assured market for milk, which has made diversification less necessary. It has also increased rapidly over the past decade, for there is evidence from both Scotland and England and Wales that a higher proportion of those abandoning dairying are not specialists but have some other livestock or cash crop as their main enterprise. Those remaining in dairying are affected by the failure of farm-gate prices to keep pace with rising costs. As the volume of milk increases, the prices obtained by farmers are correspondingly depressed below the guaranteed price; for example, the realised price in the Scottish Milk Marketing Board area increased by only 0·26 pence per gallon between 1956–7 and 1967–8, despite an increase in the guaranteed price of 1·31 pence. Individual dairy farmers can alleviate such a situation by increasing the size of their herds, whether by enlarging their farms or by better grassland management and higher stocking rates, and so take advantage of greater labour productivity; but for dairy farmers as a whole, such increases can only depress realised prices still further, unless standard quantities are increased. There seems little prospect that consumption of liquid milk will rise as fast as production and, on American experience, it may well fall as living standards improve; for per capita milk consumption in America, once higher than in Great Britain, is now only four-fifths as high and is falling at the rate of some 3 per cent. per annum.

The choice of dairying as an enterprise is thus conditioned by the possible alternatives in each region, by the improvements in labour productivity with increasing herd size and by the extent to which technical developments in milk production exceed those in other enterprises. In eastern counties, cropping has generally been more profitable and dairying has declined; on grassland areas in the western lowlands, where cropping is more difficult, dairy cattle produce a higher net income per acre than other grazing stock, but in the upland margins, with conditions less favourable for dairying, dairy cattle are losing ground to beef cattle and sheep.

Dairying must not, however, be viewed in isolation. Although beef cattle are not an important enterprise on dairy farms, the dairy herd is nevertheless the main source of young stock for beef production, and official justification for encouraging a further expansion of dairying is that this is necessary to secure an enlargement of the beef herd. The next chapter examines the complementary enterprise of beef production.

FURTHER READING

F. A. Barnes, 'The evolution of the salient patterns of milk production in England and Wales', *Trans. Inst. Brit. Geogr.* 25 (1958)

J. D. W. McQueen, 'Milk surpluses in Scotland' *Scot. Geogr. Mag.* 77 (1961)

E. S. Simpson, 'Milk production in England and Wales', *Geogr. Rev.* 49 (1959)

Milk Marketing Board, *The Structure of Dairy Farming in England and Wales 1963–4* (1965)

Ibid. *Changes in Milk Output 1963–7* (1968)

Ibid. *Dairy Herd (1965) Census* (1966)

Scottish Milk Marketing Boards, *The Changing Structure of Scottish Milk Production* (1965)

Ministry of Agriculture, *Report of the Committee of Inquiry in England and Wales* (1969)

CHAPTER 9

Beef Cattle

Fat cattle and calves are the second most important product of British agriculture, accounting for about 16 per cent. of farm sales and 23 per cent. of those from livestock and livestock products, proportions which have changed relatively little for thirty years. Governments have stimulated beef production since the Second World War through price guarantees and subsidies, and home production has risen by over 50 per cent. This increase has helped to reduce imports, which come mainly from Argentina and Australasia, and British farmers now supply over 70 per cent. of home requirements of beef and veal, compared with 49 per cent. before the Second World War. Consumption per head is lower than pre-war, largely because of competition from pig and poultry meat (see Chapter 11), but beef remains prominent in British diets, notably in Scotland, where 10·8 lb. per head was consumed in 1964, compared with an average of 8·5 lb. for Great Britain.

The present geography of beef production, like that of dairying, has thus been modified by government intervention but, for various reasons, some statistical and some arising from the nature of the beef industry, it is less easy to study. Although there are enterprise studies for various parts of the country, they are rarely contemporaneous and are often on different bases. No comprehensive surveys have been made of the whole country and there have been no marketing boards to supply the kind of information that the milk marketing boards have provided about dairying, although the recently established Meat and Livestock Commission should remedy this. Nor is the information from official statistics very helpful; in England and Wales, young herd replacements cannot be distinguished from other female beef cattle and, even in Scotland, where all categories of beef cattle are enumerated, the various ages at which they are fattened make discrimination between store and fat stock by ages somewhat arbitrary; yet this distinction is important in the geography of beef cattle, for the earlier and later stages of beef production have different distributions.

A further difficulty is that beef production is much less likely than dairying to be *the* major enterprise on farms. Beef cattle are kept in relatively small numbers on farms in all parts of the country, often as one of several minor enterprises associated with cash cropping or dairying. There are several reasons for this situation. Beef production is inextricably linked with dairying, which supplies some three-quarters of the home-bred cattle slaughtered for beef, so that most beef animals are a by-product of dairying, whether culled cows and bulls from the dairy herd, or surplus calves. In the past, beef cattle were rarely able to compete with more intensive enterprises for good land in the lowlands, but they do provide one way of cashing surplus grazing, arable by-products or other unused resources. Since they can be sold and purchased at any age from a few days old, they can fit at some stage in their lives into a great variety of farming systems, and farmers' decisions to buy or sell beef cattle may be influenced by short-term fluctuations in prices.

Tradition and other non-economic considerations strongly influence the decision to keep beef cattle, and economic factors have probably counted for less in beef production than in any other major branch of farming. Many farmers regard cattle as essential in any proper farming system; they also get pleasure out of keeping cattle, especially from showing pedigree stock, and W. H. Long has suggested that many farmers prefer this kind of recreation from their stock and will forgo profit to follow the branch of farming which appeals to them most. In any case, their profitability is often not accurately known. Formerly, beef cattle were often kept as much for their indirect contribution to farm income, particularly by providing manure, as for their sale value. Beef cattle may also graze land jointly with other livestock or use arable by-products which would otherwise be wasted, so that their feed cannot easily be costed; the practice, too, of judging stores by eye rather than by weight may lead a farmer to pay more per cwt. for store cattle than they will fetch after fattening. A farmer, especially if he is elderly, may also prefer to follow a system of farming, like rearing beef cattle, which brings less profit, but involves less effort. Beef cattle may thus be found in many places where they could hardly be justified on strict economic logic. They are also widespread because the complexities of farming systems and the variety of physical conditions provide many appropriate niches into which they can be fitted.

In addition, unlike dairying, where rearing replacements is subordinate to the main business of producing milk and generally takes place on the same farm, beef production comprises distinct stages of breeding, rearing and fattening, practised at various levels of intensity and on different farms, and often concentrated in different parts of Great Britain, although the patterns are far from clear-cut. There are therefore complex movements of beef cattle at various stages of development, some between farms in the same locality and some between different parts of the country, as with the movement of calves from dairy herds in western England to eastern counties and even Scotland. Store cattle also come from the Republic of Ireland for fattening on British farms, and all these moves have complex seasonal components, reflecting seasonal differences in the availability of feed, the period of calving and prices of store stock and fat cattle.

It is thus impossible to produce a satisfactory map of the distribution of beef cattle as a whole, but they are found in all areas except the poorest rough grazings. These apart, they are fewest in the main dairying areas, and in the intensive cropping areas of eastern England. Beef cattle are relatively most important in Scotland, where beef accounts for 24 per cent. of farm sales, compared with 13 per cent. in England and Wales, although even here it is only on the upland margins that labour requirements of beef cattle account for as much as a third of all man-days; proportions are lowest in those areas where dairying and intensive cropping are the chief enterprises (Figure 20).

An important feature of beef production is that beef cattle provide both intermediate and final products, stores and fat cattle respectively, although only the latter are included in the percentages of sales quoted above. The main aim is the production of fat bullocks and heifers, which account for over two-thirds of the cattle slaughtered. Cows and bulls from both dairy and beef herds comprise a further fifth, but they are not eligible for fat cattle guaranteed prices and are sold mainly for manufacturing. The remainder consists of calves, mainly of dairy breeds, which are slaughtered shortly after birth and used for manufacture or reared for veal, although veal is of little importance in Great Britain, accounting for only about 1 per cent. of sales of beef and veal. Many farmers, however, depend on sales of store stock, which were valued at 81 per cent. of sales of cattle from northern livestock farms included in the Farm Management Survey in 1963. Reconciling the interests of producers of store stock and those of fat cattle (in so far as they are different farmers) presents major problems. The price of store cattle fluctuates greatly and buyers of stores for fattening want it as low as possible (consistent with an adequate supply), for the price of stores is a major determinant of the profitability of fattening. Many of the store cattle, too, are raised on land of low productivity where few other enterprises are possible, and subsidies are given for keeping breeding stock and on calves intended for beef to help those who cannot fatten the cattle they rear.

The life span of beef animals ranges from under one to three years; the lower limit applies only to cattle reared under intensive systems. For other beef cattle, a comparatively long period of rearing, generally lasting one to two years, is followed by fattening for up to six months, and breeding, rearing and fattening can thus be distinguished as separate activities, each with rather different requirements and tending therefore to be located in different parts of the country. There are many permutations, for some calves intended for beef are sold within a few days of birth, while others, estimated at a quarter in Scotland, are fattened on the farms on which they were born. Nevertheless, many spend their lives on at least two farms, being bred and reared on one and fattened on another. In the past, when most cattle were slaughtered at ages exceeding two years, there was characteristically a long store period, so that cattle might change hands four or five times; but in recent years there has been a marked tendency for beef cattle to be killed earlier, so that a separate stage where cattle are bought and sold as stores has become less common, although much depends on the environment, on the breed of cattle and on the other enterprises with which beef production is associated.

BREEDING & REARING

Home-bred cattle intended for beef, i.e. excluding Irish stores, now come from three main sources: from dairy cows, the main source; from hardy beef cows kept on the upland margins; and from beef cows on lowland farms. Before specialised beef and dairy breeds emerged in the eighteenth century, beef was largely obtained from dual-purpose cattle bred in the remote uplands, whence mature animals, strong enough to make the journey, were driven along the drove roads for fattening either on grass or on arable farms near the main markets, especially London. With the rise of dairying and specialised dairy breeds, the contribution of dairy cattle to beef supplies tended to decline, for the main aim of calving on dairy farms was to maintain the milk supply; dairy farmers were primarily interested in milk production and surplus calves, which were generally unsuitable for beef and which would use grazing required for milking cattle, were slaughtered soon after birth. For two decades, however, there has been strong government encouragement to increase supplies of beef and a calf subsidy, introduced in 1946 and now £11 5s. a head for steer calves and £9 a head for heifers, encourages the retention of calves suitable for beef production. At the same time, the growing popularity of Friesian cattle and the use of artificial insemination to produce cross-bred cattle from dairy cows has increased the proportion of suitable calves from herds kept mainly for milk. These developments have induced a rapid expansion in the number of calves retained; thus, while total cattle increased by 23 per cent. between 1948 and 1965 and female calves under one year old increased in similar proportion, male cattle under one year almost trebled.

The dairy herd contributes both pure-bred and cross-bred calves for beef production. The pure dairy breeds, such as the Ayrshire, Guernsey and Jersey cattle, are generally unsuitable for beef, even if crossed with a beef bull, although the recent introduction of Charollais bulls may change this. In 1962 only about a fifth of the calves of such breeds, excluding these required as herd replacements, were retained for beef and these contributed only about 5 per cent. of all the calves retained. In contrast, four-fifths of calves of dual-purpose breeds, such as the Friesian and the Dairy Shorthorn, which were not required for breeding were retained for beef and provided almost two-thirds of all the calves retained; Friesian calves, particularly, have proved very satisfactory for systems of intensive rearing and fattening. Many of these calves of dual-purpose breeds are pure bred, but an increasing proportion, estimated at 47 per cent. in 1962, consists of crosses from a beef bull and a dairy cow. The change has largely been made possible by the increasing use of artificial insemination. In 1946 only some 64,000 cows were inseminated, but the number is now about 2 million, and the proportion of inseminations from beef breeds has increased from 15 per cent. in 1951 to 35 per cent. in 1965.

Most of these calves, whether pure-bred or beef-cross, originate on farms in the principal dairying areas, but regional differences in the proportion of the calves born to dairy cows and retained for beef cannot be determined from available data. Some indication is provided by the breed composition of the dairy herds; thus the predominance of Ayrshire cows in the Scottish dairy herd (78 per cent. in 1965) implies that few of its surplus calves are used for beef, and the high proportion of cows in South-east and Southern England which are of Ayrshire or Channel Island breeds (43 per cent. and 42 per cent., compared with an average of 26 per cent for England and Wales) suggests that dairy farms there make a smaller contribution to beef production than their share of the number of cows indicates. Although the proportion of calves from dairy herds slaughtered as calves has fallen since the Second World War and almost halved between 1957 and 1966, it varies considerably from year to year, depending largely on the price of fat stock; for example, the number of calves slaughtered rose 60 per cent. between 1965, when fat stock prices were high and many more calves were retained, and 1967, when the market had weakened.

Calves born on dairy farms and intended for beef do not normally spend long on these farms, especially if they are ineligible for the calf subsidy, for they compete for grazing or fodder with the milking herd which provides a much higher gross margin per acre. A Milk Marketing Board survey suggested that over half the steer calves born in the main dairying areas were sold within two months, compared with a fifth in northern England. A further indication of the extent of such movement is given by the 1968 June census for England and Wales, which shows that everywhere except South-west England more than half the steer calves under one year were on holdings with no dairy cows, though over 70 per cent. of the calves retained for beef originated from dairy cows. Another indicator is the ratio of steers under one year old to all cows; it would be uniform at about 35 per cent. if the calves remained in the area in which they were born (discounting any regional differences in dates of calving or mortality), but it is lowest (under 20 per cent.) in the main dairying areas and highest (over 50 per cent.) in eastern counties. For most calves from dairy cows but intended for beef there is thus a clear distinction between breeding and subsequent stages of beef production, for they soon move to other farms and other regions.

A very different situation prevails with the one-third of calves retained for beef which are born to cows kept mainly for beef production, for these calves are kept at least six months before sale as stores and sometimes much longer, depending on when they are born, on the prevailing prices for store cattle and on the feed available; where feed is adequate, as on most lowland farms with beef breeding herds, calves are generally fattened on the farms on which they were born. Such beef cows, now nearly a quarter of the cow population of Great Britain, are generally crosses of one of the eight recognised beef breeds (Aberdeen Angus, Beef Shorthorn, Devon, Galloway, Hereford, Highland, Lincoln Red and Sussex). Their number has risen sharply since the War, encouraged by subsidies for keeping hill cows and for retaining suitable calves, and increased by over 80 per cent. between 1953 and 1968, compared with a 5 per cent. increase in dairy cows in the same period.

Such beef cows are widely distributed, though numbers are low in northern Scotland where conditions are generally unfavourable for farming of any kind, and in districts devoted to intensive cropping in eastern England. Four areas, which include both upland and adjacent lowland, stand out, each having about an eighth of the breeding herd: north-east Scotland; north-east England and south-east Scotland; the Welsh borderland, especially the counties of Brecon, Hereford and Radnor; and the south-west peninsula (Figure 13). Proportionately twice as many holdings have beef cows in Scotland as in England and Wales. The percentage of holdings with beef cows varies directly with the number of beef cows in each region, ranging from 8 in Eastern England to 53 in North-east Scotland. Large herds are also more common in Scotland, the average for full-time holdings being 20 cows, compared with 11 in England and Wales; 42 per cent. of beef cows are in herds of 50 or more cows and only 11 per cent. in herds of under 10, compared with 25 per cent. and 21 per cent. respectively in England and Wales. Breeds of bull also differ. In Scotland the Aberdeen Angus and, to a lesser extent, the Beef Shorthorn predominate, although the Hereford is gaining popularity. The Aberdeen Angus and Hereford are the most important breeds in England and Wales, but there are considerable regional differences and other breeds are often popular in their name localities: thus Herefords predominate in the Welsh borderland, the Devon in the south-west and the Lincoln Red around the Wash, while the Welsh Black, a dual-purpose breed, is important in North Wales. These breeds differ in hardiness, size and rate of growth, e.g. the Hereford grows more rapidly (like other red breeds) and can use poor grazing, but does not do well in severe climatic conditions, while the Highland breed is hardy but slow growing.

Beef cows are kept in a wide range of environments. Farmers in the lowlands, wary of the variable cost of purchased stores, have increasingly acquired breeding herds to raise their own calves for fattening, with, however, the disadvantage that the breeding stock requires more than half the pasture, which could otherwise carry additional stores. Nevertheless, single-suckled herds seem to be more profitable than the traditional system of fattening purchased stores, except when store prices are low. In recent years, the time of calving in such herds has gradually been pushed back from spring to winter and even autumn so that calves can be finished the following winter. Keeping beef cows on grassland in the lowlands is most common in areas of traditional fattening on grass, such as the Lincoln marshes, coastal Northumberland, the plain of Hereford and eastern Scotland. A few beef cows may also be kept on dairy farms and on holdings where shortage of labour or capital makes extensive

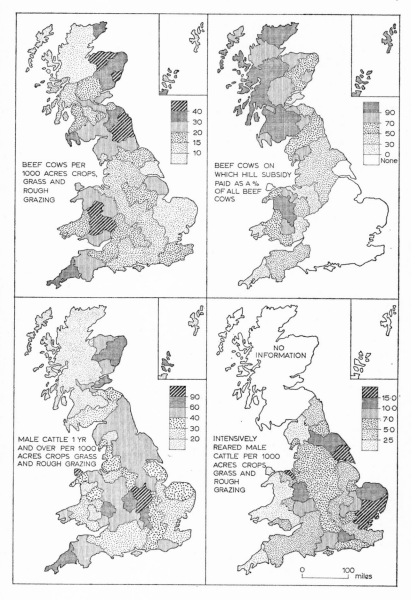

Figure 13. Beef cattle
Source: agricultural censuses and subsidy records 1965

systems necessary; they are also found on poorer land in the lowlands, and the Beef Cow subsidy, introduced in 1966 and now £8 10s. per breeding cow, provides an incentive to keep beef herds where the stocking rate does not exceed one cow to two acres. Other beef cows are on arable farms with unploughable grass or rotational leys, and tend to calve either in the autumn or in spring. Still others are in pedigree herds or are kept on home or hobby farms where commercial motives may not be important.

Keeping cows to produce calves for beef is also common on the upland margins and on the lower rough grazings; their occurrence here is much more localised, for they occupy a narrow zone between the high uplands, the domain of sheep, and the lowlands, where they tend to be displaced by dairying and other more intensive enterprises. These areas traditionally rear beef cattle and before the Second World War most herds of breeding cows were here and not scattered throughout the lowlands as at present. The land is too poor for fattening, however, and farmers depend on the sale of weaned calves or older stores. On the highest and poorest land, the characteristic breeds are the Galloway, Highland or Welsh Black, but on better ground Shorthorn and Hereford or Angus crosses may be common. Cows may be either outwintered, as in parts of Scotland, or housed, and generally calve in spring, the calves being sold in the autumn, when they are 6 to 8 months old, or as older stores at 12 or 18 months. These are lower ages than formerly, when, in Wales at least, the 3-year-old store, or runt, was the traditional product of the hills.

Winter feed constitutes the most important variable cost on such farms and is often the limiting factor to the number of beef cows which can be kept. In spring, when calving usually takes place, beef cows compete with sheep for available grazing, but these livestock also complement each other and the increase in beef cows on the hills has improved grazing and carrying capacity. Farms in these areas are generally small and dependent on family labour, much of it unpaid, and such a labour force can perhaps better provide the kind of attention breeding stock needs; quality of stockmanship is certainly a major factor in the successful keeping of beef cattle. Breeding and rearing of beef cattle is an extensive enterprise, well suited to the lower-rented land on the upland margins and so releasing lowland pastures for more remunerative enterprises; for while hardy beef cattle are much less efficient converters of feed than are dairy cattle (which tended to replace beef cattle on better land in the 1930s and 1940s), they can use more roughage and are less affected by climatic set-backs, and so are better fitted to use these poorer grazings.

The farmer keeping beef cows on the hill margins is much more the hostage of weather than his lowland counterpart, nor does he sell a product with a guaranteed price. In recognition of this, a Hill Cow subsidy was introduced in Scotland in 1941 and in England and Wales in 1943. These subsidies are payable on cows in regular breeding herds maintained on hill land not capable of being used for dairying, cash cropping or livestock fattening, even after improvement; the current subsidy is £15 15s. a head, with a maximum stocking rate of 1 cow to 4 acres. Subsidies were paid on some 450,000 cows in 1965, or 52 per cent. of the breeding herd, but the proportion varied greatly, from 83 per cent. in Scotland and 71 per cent. in Wales to 24 per cent. in England. Hill cow and calf subsidies on such farms may nearly equal the revenue from sales of cattle and may even exceed the gross margin. Figure 13, showing the proportion of breeding cows throughout Great Britain on which Hill Cow subsidy was paid in 1965, gives a relative measure of the importance of breeding herds on hill land; it reveals a contrast between the Scottish uplands, and the south-west peninsula, Wales, and northern England, which is a common feature of maps of sheep and cattle, reflecting the differences in the upland environments. Unlike breeding ewes on hill land, there is no substantial movement of beef cows to the lowlands for subsequent breeding.

An interesting study of the movements of a sample of calves born in Great Britain between 1st October 1962 and 30th September 1963 has recently been published. Of those calves, 16 per cent. were sold for immediate slaughter and 20 per cent. for rearing, while a further 4 per cent. died or were culled. Most of the calves which were purchased had been bought from elsewhere in the same region (72 per cent.), but the proportion fell to 50 per cent. or under in eastern and south-eastern England. A map, recording inter-regional movements of over

3,000 calves, shows that the main dairying areas are the principal regions supplying calves to other regions, and northern, eastern and south-eastern England the principal purchasing regions. Further, only 22 per cent. of calves moved 50 miles or more and journeys from dealers' premises were on average much longer than those from other farms.

FATTENING

Except on farms where home-bred cattle are fattened, beef cattle are sold as calves or store cattle of varying ages, to be fattened elsewhere. There is thus a general easterly movement of beef cattle, from poorer land to better, from upland to lowland, from western lowland to eastern and from Ireland to the eastern lowlands of Great Britain. These beef cattle are fattened in units of very varied sizes and under many different systems, some of which are described in the Ministry of Agriculture's bulletin, *Beef Production*, and others, with case studies, in a bulletin, *Developments in Beef Production*, by its Agricultural Economics Division; but there is little information about their distribution or relative importance, although certainly the greatest variety is found on arable farms. Seasonal changes in prices of store and fat cattle, which are lowest in autumn and highest in spring, influence the choice of systems, as do the availability of feed, labour, buildings and other resources, and the attitudes of farmers towards intensive systems.

Traditionally cattle, purchased as forward stores, have been fattened on grass in summer and on arable crops in yards in winter. These methods may be combined on the same farm, but summer fattening on grass has long characterised certain districts of high quality permanent pasture, such as the Welland Valley in Leicestershire and Northamptonshire, while fattening in yards in winter has been typical of the main arable counties. These systems persist, but younger cattle may now be purchased, and grazing and feeding in yards combined, with cattle being wintered in yards and finished on grass in spring, when liveweight gains on grass are greatest, or kept on grass in summer and fattened in yards in winter. Cattle are also fattened for early slaughter under intensive systems. The trend to earlier slaughter is clear and reflects the demand for smaller joints and leaner meat, as well as pressure on profit margins; for a shorter life provides higher margins per acre. While numbers of beef cows and younger beef cattle have increased steadily since the war, other cattle of two years old and over, the age at which fat cattle have traditionally been slaughtered, decreased by 44 per cent. between 1953 and 1967.

Figure 13, showing the distribution of other male cattle of one year old and over, indicates the location of beef fattening, although, being a map of summer distribution, it emphasises the importance of fattening on grass; the main arable areas would be more prominent on a map based on the December census. With a few exceptions, such as Anglesey and south-west England, the principal fattening areas are located on the eastern side of Great Britain and include two of those shown as important for breeding in Figure 13.

The leading area lies in the three counties of Leicester, Northampton and Rutland. These and other areas such as north-east Northumberland are traditional fattening areas where standards of stockmanship and grassland management are high; for example, liveweight gains on Midland pastures are twice as high as those in Wales. Grass is certainly the cheapest feed, and labour costs are lower than with yarded cattle; it is even claimed that the best quality pastures give a better return from beef than from dairying. Only sheep graze the fattening pastures in winter and cattle numbers slowly increase until May, when a stocking rate of between $1\frac{1}{4}$ and $1\frac{1}{2}$ beasts per acre is achieved. The best land is reserved for bullocks and the first cattle are marketed in June before prices fall: delay in marketing may well lead to a sharp fall in profits. A second batch may be finished in late summer and early autumn, when liveweight gains are rising again. Older cattle still predominate in these areas; the Midland pastures are unsuitable for younger cattle, which cannot get a sufficient intake of nutrients on such grass.

Beef cattle are fattened elsewhere on grass, although they cannot normally give a sufficiently high return to compete with dairy cattle as a main enterprise; however, they need little other than grazing, water and fencing, and require much less labour than dairy cattle. Grass management must be of a high order if such fattening on grass is to pay, and it is management and stockmanship which largely explain the continued importance of these traditional fattening pastures. The price of store steers and the costs of wintering are also critical elements in determining the profitability of winter fattening.

Feeding cattle in yards on arable crops and residues was an essential element in the Norfolk 4-course husbandry and many farmers still regard the making of manure as the prime justification for keeping cattle. Forward stores of two years and over, purchased in the autumn, were fed on roots and concentrates, and usually two batches were fattened. Younger cattle are now bought and may spend the summer on surplus grass; alternatively they may be home-bred. They are fattened in yards in winter on arable by-products such as sugar beet pulp and silage made from pea haulms, and only a small acreage of fodder roots is now grown. The justification for the system in present circumstances is that it employs by-products or buildings which would not otherwise be used, and uses labour in the arable farmer's slackest period.

Both fattening on grass and fattening on arable farms have become more intensive as rents have risen and profit margins have been squeezed. This trend reaches its most extreme form in the production of baby beef from calves, often Friesians, purchased soon after birth and fed on a diet of concentrates to reach slaughter weight at twelve months or less. Food conversion is most efficient in the early stages of growth and a high rate of livestock gain is maintained. Spring-born calves fattened in this way reach the market when prices are highest and labour requirements are less than under traditional systems of yard fattening; this system also produces the lean beef and smaller joints required by the market. Critical considerations are buying prices for calves, mortality among calves and cost of feed. Intensive systems are commonest in eastern counties (Figure 13), where cereals are available and farmers may be able to reduce costs by milling their own grain. In 1965, 8 per cent. of the male cattle under 2 years in England and Wales were kept under intensive systems and in Scotland 4 per cent. of the fat cattle were produced under intensive systems in 1964, although increases in the price of suitable calves, which more than doubled in 1961 and 1964, have lessened enthusiasm for barley beef.

BEEF CATTLE & FARMING SYSTEMS

The place of beef cattle in farming systems is thus very varied and there may even be different systems on the same farm. The choice is probably greatest for arable farmers and least in the uplands, where few alternatives of either method or enterprises are possible. Generally, the importance of breeding declines and of fattening increases in moving from less to more intensive types of farming; for example, the ratio of other beef cattle to beef cows in Scotland increases from 1·2 to 1 on hill sheep farms, through 2·8 to 1 on rearing with arable farms, to 6·2 to 1 on cropping farms. Tables 16 and 17 show the proportion of beef cows, representing one of the first stages in beef production, on holdings of different kinds in England and Wales and in Scotland respectively. Over half the beef cows in England and Wales were on livestock holdings of various kinds, but 13 per cent. were on cropping farms (nearly a quarter of which had beef cows) and 5 per cent were on dairy farms, where herds averaged only four cows. Even intensive pig and poultry farms and horticultural holdings had some beef cows. In Scotland beef cows are kept mainly on upland and mixed farms (rearing with arable, rearing with intensive livestock and arable with feeding), and upland farms have the largest herds. Although hill sheep farms had only 7 per cent. of the cows, beef cows were kept on three-quarters of them.

Such measures of the distribution of beef enterprises among farms of different types do not adequately show the importance of beef cattle in those farming systems. In England and Wales, the contribution of cattle to farm sales is highest on livestock farms, averaging 31 per cent. on all types and reaching 54 per cent. on livestock fattening farms, and declining to 10 per cent. on dairy farms and 8 per cent. on arable farms. The situation in Scotland is similar, except that cattle contribute the largest share (39 per cent.) on cropping farms. In England and Wales livestock farms account for the largest proportion of beef cows in all regions except Eastern England and the East Midlands, where 70 per cent. and 44 per cent of the beef cows respectively were on crop farms (Table 16). For other male cattle of 1 year old and over, which include both animals being fattened and older stores, livestock farms are again the leading category, but the contribution of cropping farms in the Eastern Region, the East Midlands and Yorkshire and Lancashire is higher, as is

TABLE 16
BEEF COWS AND OTHER MALE CATTLE IN EACH REGION OF ENGLAND AND IN WALES, 1966

	England and Wales	Eastern	South-east	East Midlands	West Midlands	South-west	Yorks and Lancs	Northern	Wales
Beef Cows									
Proportion of beef cows by type of holding									
All livestock types	62	13	39	36	63	64	58	82	90
Mixed	11	10	22	12	13	18	12	8	+
Dairying	+	+	+	+	+	11	10	+	+
Cropping	14	70	22	44	11	+	16	+	+
Percentage of beef cows in England and Wales									
	100	6	8	9	12	18	5	20	22
Percentage of holding with beef cows									
	19	8	12	17	19	21	13	30	33
Other Male Cattle 1-year-old and over									
Proportion of other male cattle 1-year-old by type of holding									
All livestock types	41	17	41	42	42	42	24	47	64
Mixed	17	10	19	13	18	25	17	18	10
Dairying	15	+	14	10	17	23	14	14	22
Cropping	24	62	19	33	+	+	39	19	+
Percentage of male cattle 1-year-old and over in England and Wales									
	100	9	10	15	12	17	10	15	13
Percentage of holdings with other male cattle 1-year-old and over									
	29	15	17	35	29	33	26	47	40
Other Male Cattle under 1-year-old									
Proportion of other male cattle under 1-year-old by type of holding									
All livestock types	35	11	28	24	33	37	23	40	65
Mixed	18	18	21	16	19	25	17	17	+
Dairying	24	+	25	23	30	31	24	24	23
Cropping	20	59	19	31	14	+	31	12	+
Percentage of male cattle under 1-year-old in England and Wales									
	100	10	10	10	13	16	10	16	14
Percentage of holdings with male cattle under 1 year									
	34	16	23	34	33	36	30	55	47

Source: Ministry of Agriculture
+ = less than 10 per cent.

their share of older cattle. The association of other male cattle of under one year old (which include both calves and young store cattle) with farm type is very similar, and it is not surprising, in view of the contribution of calves from dairy herds and the movement of calves to farms in eastern England, that other male cattle under one year old are found on almost twice as many holdings as beef cows and are more uniformly distributed.

The value of such an analysis is limited by the size and heterogeneity of the regions, by the fact that store cattle cannot be distinguished in census data from those being fattened and by the grouping of all livestock farms irrespective of their end products. A more illuminating analysis is possible for Scotland, where studies of the structure of the industry have been made by agricultural economists and the statistical regions are more homogeneous than in England.

The beef breeding herd in Scotland is proportionately much larger than that in England and Wales and there are now slightly more beef cows than dairy cows. Furthermore, since most of the dairy cows are Ayrshires, their contribution to the number of fat cattle slaughtered is small and probably consists mainly of Friesians and of cross-bred cattle; it was estimated in 1964 at some 82,000, or 17 per cent. of all fat cattle, including 12,000 reared intensively. Some 60–70,000 beef calves from Scottish herds went to England, but about 100,000 dairy calves of up to a few weeks old, mainly Friesian and cross-bred calves, entered Scotland from England, a net gain of 30–40,000. Irish stores contributed 95,000 fat cattle, about 20 per cent. of the fat cattle slaughtered, and cows and bulls, 90 per cent. of them from the dairy herd, contributed a fifth of total beef supplies. The dairy herd thus provided some 31 per cent of beef produced in Scotland.

Within Scotland, beef cattle are commonest in the North-east (Table 17), which has nearly two-fifths and where 72 per cent. of holdings have some beef cattle; the concentration of 'other beef cattle' in the North-east is even more striking. The distribution of the breeding herd is similar, although the predominance of the North-east is less marked and fewer holdings have breeding cows. Breeding herds are generally smaller in the Highlands and the North-east, with 52 per cent. and 45 per cent. respectively of the cows in herds of less than 30 cows, while large herds are commoner in the East-central

TABLE 17
BEEF CATTLE IN SCOTLAND, 1967

	Highland	North-east	East-central	South-east	South-west	Scotland
Beef cattle: numbers and holdings, 1967						
% in each region	13	38	15	11	24	100
% of holdings with beef cattle	44	72	58	58	59	57
Beef cows: numbers and holdings, 1967						
% in each region	19	31	16	12	22	100
% of holdings with beef cows	34	53	33	33	27	38
Beef cows: percentage of cows in each herd size group, 1967						
Herd size:						
−10 cows	23	14	4	3	5	11
10–29	29	31	20	15	22	25
30–49	16	20	26	24	27	22
50–69	12	12	18	22	20	16
70–	20	24	32	36	27	26
ALL	100	100	100	100	100	100
Other beef cattle 6 months old and under 1 year, 1967						
% in each region	8	32	16	12	32	100
% of holdings with other male cattle 6–12 months	11	29	24	25	33	23
Other beef cattle 1-year-old and under 2 years, 1967						
% in each region	7	46	15	9	22	100
% of holdings with other male cattle 1–2	25	58	41	42	44	41
Other beef cattle 2 years old and over, 1967						
% in each region	6	48	17	9	21	100
% of holdings with other male cattle 2 and over	8	22	13	14	16	15
Calves and store cattle: sales, 1961						
% of young calves sold	4	17	4	3	72	100
% of store cattle sold	13	33	18	11	25	100

Source: Department of Agriculture and *Scottish Agricultural Economics*

and South-east regions, with 50 per cent. and 58 per cent. of the cows in herds with 50 or more cows. Three-quarters of these beef cows are on upland and hill farms and attract Hill Cow subsidy (Figure 13), so that it is not surprising that only a quarter of the fat cattle were home-bred (Table 18).

Calves from England and stores from Ireland went mainly to the East-central and North-east (43 and 52 per cent. respectively). The South-west, where dairy cattle predominate, provided most of the young calves sold, but it also had the largest share of the dairy cattle fattened for beef and was the only region where more than half the fat cattle slaughtered were of dairy breeds. Apart from the Highlands, where most of the cattle fattened were home-bred and conditions are generally harshest, the regions contributed to the supply of store cattle in roughly the same proportion as their share of breeding stock.

Sources of stores for fattening varied from region to region (Table 18). The East and South-east depended on Irish cattle for 36 per cent. and 40 per cent. of their stores respectively, while both the North-east and the South-east made relatively large purchases (13 per cent. and 18 per cent. respectively) of cattle under 2 months old mainly from dairy herds; the South-west, in contrast, purchased mainly cattle over 1-year-old, presumably from dairy farms in the same region. Few stores of beef breeds were sold before weaning, whereas those of dairy breeds tended to be sold either very young or at over 1 year.

There is a similar contrast in the age of slaughter. Beef breeds, especially if fed in yards, mature more rapidly than dairy breeds, more of which were slaughtered at ages exceeding 30 months (17 per cent. compared with 8 per cent.); but dairy calves are more frequently reared intensively and slaughtered at under 15 months (8·5 per cent. compared with 2 per cent.). The great majority of fat cattle, however, were slaughtered at between 21 and 30 months, the South-east having the highest proportion of those slaughtered at under 15 months (8 per cent.) and the North-east that of cattle slaughtered at over 30 months (14 per cent.). It is noteworthy that a small minority of fattening farms, numbering little more than a thousand (8 per cent.), provided nearly half the cattle sold (46 per cent.) in 1961, while 8,000 farms (59 per cent.) from which fewer than 20 head were sold, provided only 12 per cent.

TABLE 18
FAT CATTLE IN SCOTLAND, 1963

	Highland	North-east	East central	South-east	South-west	Scotland
Percentage of Dairy and Beef Breeds, 1963						
% of beef breeds in each region	3	43	29	14	13	100
% of dairy breeds in each region	2	28	9	11	49	100
Beef breeds as % of regional herd	80	82	90	78	43	75
Percentage of Fat Cattle from different sources, 1963*						
Home-bred	43	19	19	29	34	24
Other Scotland:						
−2 months		9	4	12	7	8
2–12 months	58	30	23	6	11	20
12+ months		29	25	19	41	30
English calves	—	4	1	6	3	3
Irish stores	—	9	29	28	5	15
Percentage of Fat Cattle fed under different systems, 1963*						
Intensive	2	3	2	8	2	4
Court fed and finished		17	24	47	20	24
Court fed and grass finished	98	30	21	12	15	22
Grass fed and court finished		37	35	25	26	32
Grass fed and finished		13	18	8	37	18
Percentage of Gross Output from fat cattle and calves, 1961						
% of gross output for Scotland	4	39	25	14	18	100
% of gross output in each region	10	36	25	22	14	23

Source: Scottish Agricultural Economics
* full-time holdings only

There were also considerable differences in the systems of feeding adopted throughout Scotland, with intensive systems accounting for the largest share (8 per cent.) in the South-east.

The clearest regional contrast is between grass feeding and yard (court) feeding, 37 per cent. of the cattle fattened in the South-west being fed and finished on grass, and 47 per cent. of those in the South-east being fed and finished in yards; it is not surprising that beef breeds are yard-fattened in one of the chief arable regions and dairy breeds grass fattened in the main grassland region.

The pattern of beef production in Scotland shows clear affinities with the kind of feed available and with the suitability of the various regions for other enterprises, of which the contribution of fat cattle to gross output in Table 18 is an indication. The North-east, where conditions are not very suitable for cash cropping or for liquid milk production, has both the highest proportion of its regional output from fat cattle and calves, and makes the largest contributions to Scottish production, whereas the Highlands, where most of the land is suitable for only very extensive systems of farming, has the smallest contribution, not because of competition from other enterprises, but because of the general poverty of the land.

Despite this wealth of information on Scotland, it is still difficult to provide a satisfying explanation of the location of beef production, partly because of lack of regional economic data for each enterprise and partly because of the variety of motives which lead farmers to rear and fatten beef cattle. We should perhaps remember the conclusion of a N.A.A.S. survey of beef herds in Northumberland, which sought to discover why some farmers made more from single-suckled calves than others and found that the better farmers on better land had more money to buy better cows and better bulls, which they fed better, so the calves were bigger and brought higher prices. Not inappropriate also is the comment of a Scottish migrant farmer to eastern England in the early twentieth century, 'Cow keepers keep cows; gentlemen fatten bullocks.'

FURTHER READING

W. H. Long, 'The place of beef production in British farming', *J. Agric. Econ.* 9 (1950–2)

F. McIntosh, 'Fat cattle production in Scotland 1963–4', *Scot. Agric. Econ.* 15 (1965)

Ministry of Agriculture, *Calf Wastage and Husbandry in Britain* 1962, Annual Disease Surveys, Rept. No. 5 (1968)

CHAPTER 10

Sheep and Lambs

Sheep are the most numerous grazing animals in Great Britain and, because they are generally managed under extensive systems, they occupy a large area, including most of the one-third of the country classified as rough grazing; they also have exclusive use of more land than any other enterprise, although this is not because of their greater competitive power, but because no other enterprises are possible. In money terms, they are the least important livestock: the return per acre is low and they provide only some 6 per cent. of farm sales and only 8 per cent. of those of livestock and livestock products.

Nevertheless, these figures understate the importance of sheep in the agricultural geography of Great Britain. Their distribution is more closely related to natural conditions than is that of other livestock, which are insulated to varying degrees from the soil and climate of their immediate environment by housing or by the supply of supplementary feed. This contrast has become greater in recent years, because sheep have been less affected than other livestock by the advances in livestock feeding and the application of capital to housing and mechanisation associated with the Second Agricultural Revolution. Secondly, sheep offer a striking example of the interdependence of farming systems in different areas, for the movement of breeding and store stock provides links between farms of different kinds. In this respect, they exhibit, although more markedly, several characteristics already noted in beef cattle. Lastly, their distribution has undergone more marked changes than that of any other livestock, for while the regional importance of beef and dairy cattle, pigs and poultry has also changed, sheep alone are now negligible in areas where they were once a major element in the agricultural economy.

The main product of sheep farming is the sale of fat lambs,

for relatively little mutton is now marketed because of changes in family size and consumer preferences. Wool, once the most important product of English agriculture (as the Lord Chancellor's Woolsack illustrates), now accounts for only about 15 per cent. of farm sales, although the proportion varies regionally. These products face competition from imports, mainly from New Zealand, Australia, and Argentina, and the United Kingdom, as one of the few developed countries in which lamb and mutton constitute a large proportion of meat consumption, occupies a unique place in world trade in these commodities. It is also a major importer of wool on which its dependence on overseas supplies is even greater, although home production and imports serve different markets. Home supplies constitute c. 40–50 per cent. of requirements of lamb and mutton but only about 12 per cent. of those of wool. Fortunately for the British farmer, home production of meat and that from the southern hemisphere complement each other to some degree in that they reach the market at different times; the home product also commands a premium over the frozen imported lamb. British wools similarly serve a specialist market for carpets and tweeds, and are even exported.

THE DISTRIBUTION OF SHEEP

Sheep are primarily characteristic of the uplands and fringing lowlands (Figure 14). Only the East Midlands and Kent are important areas for sheep in the English lowlands; Romney Marsh, with the highest density of sheep per acre anywhere in Great Britain, is unique among lowland areas in that sheep are the leading livestock. Densities vary considerably in the uplands, with values highest in Wales, somewhat lower in the northern Pennines and the Southern Uplands, and lowest in the Highlands; but these are averages and stocking intensities range more widely, even within the uplands, from one acre per breeding ewe in parts of Wales to fourteen acres in north Scotland. In the lowlands sheep are least numerous in the principal arable districts of eastern England. The main census, in June, shows no major differences in the distribution of all sheep and of breeding stock.

A similar pattern appears in the ratio of sheep to other

livestock (Figure 14). Sheep are relatively most important in the uplands; but the ranking of the various hill areas is different from that on the map of distribution; sheep account for over 70 per cent. of livestock units in the Western Highlands and the Southern Uplands (the only area where sheep are both numerous and relatively important), between 50 and 70 per cent. in Wales and other Scottish uplands, and below 55 per cent. in the Pennines and south-west England, where cattle have a larger share of hill land. Some of these differences arise from the shape of the statistical units in relation to the areas of rough grazing, but the broad regional contrasts are valid.

TABLE 19
PERCENTAGE OF HOLDINGS WITH SHEEP, BY REGIONS

ENGLAND & WALES, 1968

	Eastern	South-east	East Midlands	West Midlands	South-west	Northern	Yorks and Lancs	Wales
Sheep	4	16	26	29	29	55	28	49
Breeding ewes	4	15	25	28	27	52	26	51

SCOTLAND. December, 1967

	Highland	North-east	East-central	South-east	South-west
Sheep	62	43	41	52	42
Breeding ewes	59	36	32	44	35

Source: Agricultural Departments

Marked differences also exist between farms, especially in the lowlands, for sheep are found on only 25 per cent. of the holdings in England, 51 per cent. in Wales and 46 per cent. in Scotland (Table 19). Figures for the English regions show even greater differences, ranging from 4 per cent. in the Eastern Region, the main area of arable farming, to 55 per cent. in the Northern Region; in Scotland, sheep are generally found on many more farms and the range (admittedly for December) is only from 69 per cent. in the Highlands to 41 per cent. in East Central Scotland. There are rather more holdings without breeding sheep than without any sheep, but there is little difference in the regional importance of either category. What is certain is that both the proportion of holdings without sheep,

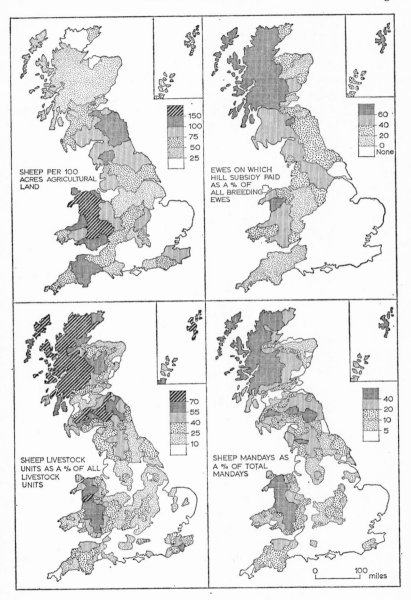

Figure 14. Sheep
Source: agricultural censuses and subsidy records 1965

which is highest on the smallest holdings, and the size of flocks have been increasing, for despite a rise in the sheep population of England and Wales, the number of sheep flocks declined by 20 per cent. between 1960 and 1968, while the proportion of sheep in flocks of 500 sheep or over rose from 9 to 49 per cent.

Sheep are a most difficult category of livestock to analyse from available statistics because of the great seasonal movements that take place and because of regional differences in the dates of lambing. In June, numbers are at their height, for although some early lambs will already have been slaughtered and some flying flocks sold, most lambs will still be on the farms on which they were born. Even so, these may be understated since some upland farmers will not have begun to count their lamb crop. Later in the year, the slaughter of most of these lambs reduces the sheep population to almost half by the time of the December census and there are complex movements of breeding stock and of lambs for fattening. In general, the hills contain only breeding stock in winter but, despite local movements from higher to lower ground, the broad pattern of distribution is similar at all seasons and the contrasts between northern and western counties and those of the south and east remain.

The distribution shown in Figure 14 should also be qualified by other maps showing different breeds of sheep. These number over 40 and vary in importance and in localisation, some, such as the Scottish Blackface and the Cheviot, being widely distributed, and others, such as the Herdwick and the Dartmoor, found in the Lake District and Dartmoor respectively, highly localised. The breeds can be broadly divided into longwoolled, large sheep with heavy fleece, which include the Leicester, the Border Leicester and the Romney Marsh; shortwoolled, including the six Down breeds, the Clun Forest and the Kerry Hill; and small, hardy, mountain breeds, of which the Blackface, Cheviot and Welsh Mountain are the most important. Except for those breeds on which hill sheep subsidy is paid, there are no accurate records of breed distributions, and estimates are complicated by the many crosses between different breeds; some of these, such as the Half-Bred, have a long tradition and almost equal status with their parent breeds. An estimate of the breed composition of the ewe flock of the United Kingdom in 1962 put the Blackface at 25 per cent.,

Welsh Mountain 15 per cent., Half-bred 15 per cent., other cross-bred ewes 10–15 per cent., and Cheviot, Shropshire, Clun, Kerry Hill, Lincoln and Masham about 5 per cent. each. Differences in the size and feed requirements of different breeds help to explain differences in density; high densities in Wales in part reflect the small size of the Welsh Mountain breed, larger numbers of which can be supported on a given area than other mountain breeds. Differences in wool, meat and fertility are also important, although to some extent they offset one another financially. Mountain breeds produce a small fleece of $2\frac{1}{2}$–5 lbs. and the wool is not of the highest quality, being used mainly for carpet-making; their small carcases, however, are excellent for modern market requirements. Long-woolled breeds have heavy fleeces, weighing as much as 16 lbs. for the South Devon, but the carcases are too large for present tastes.

FACTORS AFFECTING THE DISTRIBUTION OF SHEEP

Essentially, this distribution of sheep reflects both their ability to use upland environments where other forms of agriculture are difficult and their inability to compete with other enterprises on lowland farms. It has changed greatly over time as changing markets and technology have altered the comparative advantage of other livestock and crop enterprises; it is conditioned by regional differences in the quality of grazing and in land holding and by the labour and capital requirements of sheep. It would be wrong to assume that the present distribution reflects a perfect adjustment to existing conditions; attitudes towards sheep keeping are clearly important, epitomised by a Romney Marsh farmer who, when asked whether he would consider keeping pigs or poultry, dismissed the enquiry with the remark, 'This is a sheep farm'.

Although the housing of sheep in winter is under discussion and experiment, and sheep are housed in other European countries, a distinctive feature of sheep farming has been its small requirements of capital and specialised equipment. Fencing is needed on lowland farms, but although greater expenditure on fencing on upland farms might be justified by the higher stocking and better grazing which would result, large numbers of sheep on hill farms range over open moorland unrestrained by fencing. Expenditure on buildings is negligible, although shelter would reduce losses through exposure and make better use of available feed. Sheep do not generally need piped water, for their requirements can generally be met by moisture in their fodder. Nor is much other specialised equipment required. This independence of capital equipment is a mixed blessing for sheep keeping in the lowlands, since it implies that this enterprise can be given up easily. Despite the fact that a large proportion of sheep are kept on the nearest approach to bare unequipped land in Great Britain, rent (whether actual or notional) represented a fifth of the cost of keeping sheep under extensive systems before the Second World War and represents between 10 and 15 per cent. today.

The labour requirements of sheep are small and highly seasonal; for example, 42 per cent. of the annual labour budget on a sample of flocks producing store lambs in south-west England was used in the months of March, April and May and the labour requirements of the month of heaviest labour use was over three times that of the month with least. When much labour is required, however, it may be needed for long and irregular hours, especially at lambing. Traditionally, labour was highly skilled and the shepherd among the elite of farm workers. On hill farms in areas of long-continued depopulation, the difficulty of finding and keeping shepherds has been cited as a major restraint on sheep production, although its importance is probably exaggerated; there is no satisfactory evidence to show whether any regional differences occur within the uplands, although the situation is probably worse in the Highlands than in southern Scotland.

A flock of 500 ewes is now regarded as the minimum necessary to employ one man full-time, yet this size is attained by only 3 per cent. of the flocks in England, 6 per cent. in Wales and 7 per cent. in Scotland (although they contain respectively 22 per cent., 31 per cent. and 47 per cent. of the breeding sheep), and as many as 64 per cent. of flocks in Scotland have fewer than 100 ewes. There is no obvious pattern and no simple relationship between flock size and either the proportion of holdings with sheep or the proportion of the ewe flock in the different regions; for example, 22 per cent. of the regional flock is on holdings with 500 or more ewes in the Eastern Region, where sheep are returned on only 4 per cent. of holdings and only 1 per cent. of the national flock is to be found, while in the West Midlands the proportions are 10, 27 and 10 per cent. respectively. In general, however, the proportion is highest where sheep are most important, although even there only a bare majority of the breeding sheep are in flocks large enough to employ one man full-time; and while 8 per cent. of the breeding sheep in England and Wales and 12 per cent. in Scotland are on part-time holdings, on many other holdings sheep must share labour with other enterprises.

Sheep kept on the uplands depend very largely on the semi-natural grazing available, and differences in the vegetation largely account for differences in stocking, although breed of sheep is also important; for large breeds need a greater supply of food simply to satisfy maintenance requirements. Both the nutritional value of grazing and its availability throughout the year are important. As shown in Chapter 3, the length of the grazing season and the rate of growth decrease with elevation and these gradients are reflected in the sheep density map. The bent-fescue pastures of the lower slopes of the uplands provide the most nutritious and longest grazing in hill areas and carry the highest densities, although these better pastures have been heavily invaded by bracken during the past century. The higher uplands are mainly under purple moor grass (*molinia*), mat grass (*nardus*) or heather (*calluna*), and there are also moors dominated by sedges and sphagnum moss which provide some summer grazing. *Molinia* provides abundant feed in early summer, but little in winter; heather supplies grazing all the year round, provided it is regularly burnt; and *nardus* is the poorest grazing, being generally avoided by sheep, and supplying feed only for a brief period in summer. The lowest densities are on the moors of north Scotland, where carrying capacity is further reduced by the use of large tracts as deer forest; the heather moors similarly have to support both grouse and sheep and their respective requirements conflict to some extent. In general, grazing on the upland pastures is always inadequate, barely sufficient in winter to meet even the maintenance requirements of the breeding flock, although ewes are then carrying lambs. Furthermore, the areas of inferior grazing are those where farms include little improved land to provide supplementary feed and in the poorest areas, like the Highlands, sheep must move to low ground in winter. There is no comprehensive evidence of the extent to which sheep are fed. In a survey in Wales in the mid-1950s, four-fifths of flocks were not being regularly fed, and a third of farmers questioned even disapproved of supplementary feeding of hill sheep; but there has probably been a considerable improvement since.

Because of these problems and of the fact that hill pastures generally cannot fatten sheep, so that farmers cannot benefit directly from guaranteed prices, a hill sheep subsidy, the first subsidy limited geographically, was introduced in 1941 for ewes of hardy breeds on hill farms. The subsidy was later divided into two categories, full rate for self-replacing flocks and half-rate for those where replacements were bought. In 1964 the problem of winter feed received further recognition by a winter-keep scheme, and conditions for hill sheep subsidy were further widened in 1967 to include flocks on lower ground. The proportion of the breeding flock on which hill sheep subsidy is paid thus provides some indication of the quality of the grazing (Figure 14); in south-west England, less than 10 per cent. of ewes were subsidised in 1965, over most of Wales between 40 and 60 per cent., and in north-west Scotland over 60 per cent., reaching 82 per cent. in Sutherland. The full rate is now 21s. per ewe, which may represent a considerable proportion of net income on hill sheep farms.

The carrying capacity of lowland grazing also varies greatly, ranging from enclosed permanent pasture on the hill margins that differs little from open moorland, to Romney Marsh, where pastures can support two adult sheep per acre in winter and seven in summer; but such differences are not a major factor in explaining the varying importance of sheep in the lowlands and, in any case, sheep may feed on arable crops or their residues (such as sugar beet tops and pea haulms) and supplementary feed is widely available to those sheep kept on grass which they generally have to share, in a subordinate capacity, with cattle; for lowland farmers must either complement sheep with other livestock or intensify stocking if sheep are to be retained.

The effects of differences in carrying capacity in the uplands are moderated by variations in the way land is held, in particular by the size of holding and the proportion of common grazings, but even within the lowlands the distribution of sheep is affected by the relative importance of large and small holdings; there is a clear relationship between acreage and the percentage of holdings with sheep, which rises with size of holding (Table 20). This holds both for the Eastern Region, with under 1 per cent. of the sheep in England and Wales, and for the South-East Region, with 7 per cent. For the uplands information on holding size is available only by acreages under

TABLE 20
PERCENTAGE OF HOLDINGS WITH SHEEP

	-5	5-19¾	20-49¾	50-99¾	100-199¾	200-299¾	300-499¾	500-999¾	1000-
			Holding size (acres of crops and grass)						
			Percentage of holdings with sheep						
Eastern England	1	2	2	3	6	8	12	16	22
South-east England	3	7	13	19	24	31	36	40	53
Wales	24	31	42	59	74	81	88	89	80
England and Wales	5	13	24	38	48	51	48	44	44

Source: Ministry of Agriculture

crops and grass, although most holdings will have a substantial acreage under rough grazing, and many will have grazing rights on common land, which effectively increases their acreage. Data for Wales in Table 20 should be interpreted accordingly, but the same relationship holds in all except the high categories and almost certainly an analysis of holdings measured by total acreage would yield very similar results. Where holdings are large, sheep will be more evenly distributed than where small holdings predominate, although evaluation of the full significance of these figures would require more adequate data on holding size and on stocking densities on such holdings.

While common grazing affects such relationships, its significance lies rather in the lack of control over stocking apparently existing in many areas. Commons in both Wales and the northern Pennines, for example, are alleged to be over-stocked, and this must contribute to higher densities in such areas. In the crofting parishes of north-west Scotland a similar situation exists, although some common pastures are occupied by club stocks, a pooling of the sheep of the individual crofters, sometimes under a full-time shepherd.

The concentration of sheep in and around the uplands reflects the ability of sheep to utilise poor pasture but does not imply the suitability of these areas for sheep. Certainly sheep can better withstand low temperatures and high humidity than other stock, but survival in upland environments exacts its price. Food is poor, more energy must be expended in searching for it and more is required for maintenance; ewe lambs do not generally breed until their second season; and lambing rates are much lower and mortality rates greater than in the lowlands, not only for sheep in general, but also for sheep of the same breed. Blackface ewes in mountain environments may average 70–90 lambs per 100 ewes, but achieve 140–160 in the lowlands, and ewe lambs drafted to the lowlands may achieve lambing rates 50 per cent. above those reached in the uplands. Much depends on breed and management: Kent ewes, for example, averaged 120 lambs per 100 ewes and Scottish Half-breds 150 on Romney Marsh in 1957. In general, rates fall with altitude, and evidence from Brecon gives lambing rates of 60–65 per cent. in hills, 80–88 per cent. on marginal land and 110–150 per cent. in the lowland. Differences also exist within the uplands, depending largely on environmental conditions; Blackface flocks show percentages of 60–75 per cent. in north Argyll and 85–100 per cent. in Galloway. Percentages too, vary greatly from season to season, and lambing rates may fall by as much as 40–50 per cent. in a bad spring. These differences, of course, result partly from differences in mortality, especially among ewes in winter, but also among lambs. Mortality rates are lower in the lowlands than on the uplands and on the better uplands than on the poorer; in the Balfour Report on Hill Sheep Farming in Scotland, 70 per cent. of farmers in Lanarkshire reported ewe losses of between 1–5 per cent., and only 2 per cent. losses of 11 per cent. or over, compared with 13 per cent. and 33 per cent. respectively in Argyll. Because of such large losses, flock replacements in hill flocks in Scotland greatly exceed those drafted out.

THE CHANGING DISTRIBUTION OF SHEEP

Sheep can thus survive and breed in the uplands, although less easily than in the more favoured lowlands, but this concentration of sheep in the west and north reflects primarily their inability to compete with other enterprises in the lowlands. This situation is best considered historically. Sheep were widely kept in the lowlands in medieval England and the growing importance of wool was primarily responsible for the enclosure of arable land in Tudor times and its conversion to permanent pasture. Sheep shared the upland rough grazings with cattle, which often outnumbered them, and the pastures were grazed only in summer under various systems of transhumance. Large-scale sheep farming on the uplands was begun on the Pennines in the twelfth century by the monks of Cistercian monasteries and spread to other upland areas, reaching the Highlands only in the late eighteenth century and North Scotland in the early nineteenth century. With the development of such large-scale sheep farming, sheep came to pre-empt the use of large tracts throughout the year, and new farms were created out of the upland grazings, which were thus separated from their lowland counterparts. Such continuous grazing has adversely affected the quality of the pastures and may be responsible for a reduction in numbers of sheep in some areas. Stocking in the north of Scotland was at its highest level in the 1870s, but was affected by the creation of deer forests, which occupied 3 million acres at their peak in 1919; although much of the land was too poor for sheep or had been voluntarily abandoned, some sheep were undoubtedly displaced.

In the lowlands, sheep were kept on rough grazing or permanent pastures, but during the Agricultural Revolution, they became incorporated into various systems of mixed husbandry, notably the Norfolk 4-course rotation, using the Down breeds, which were then being developed; under these systems, practised particularly on light land, sheep were close-folded on arable crops, mainly turnips, and both fertilised and consolidated land through 'the golden hoof'. Indeed, it was widely believed that only in this way could light land be cropped, and sheep were kept on such farms as much for their manurial value as for any revenue they brought; it was said that sheep were kept to manure the land 'to grow wheat to feed the people'.

By the 1870s, sheep were distributed throughout Great Britain and densities were as high in counties such as Dorset and Lincolnshire as in Wales or the Southern Uplands. Thereafter, competition from overseas suppliers of lamb and mutton and from other livestock enterprises led to the virtual disappearance of sheep farming from many lowland areas, especially those where sheep had been folded on arable land. The general unprofitability of arable farming led farmers to lay land to grass and to reduce enterprises that were expensive of labour, such as folded sheep, and folding has thus virtually disappeared from farms in lowland England, although sheep are still fattened on roots on arable farms in Scotland and northern England. Folding in fact was not essential for the fertility of the soil, which could be maintained more cheaply in other ways. More recently, arable farmers in eastern England have increasingly tended to abandon all livestock husbandry and to remove hedges, with a consequential decline in numbers of sheep and of other grazing stock.

Although in the last quarter of the nineteenth century the numbers of sheep kept on grassland tended to increase in areas where arable land was being converted to permanent pasture, such sheep faced increasing competition from dairy cattle and numbers subsequently fell, though not as fast or as far as in the arable areas; a contributory reason around large towns may have been the growing number of dogs and increasing problems of trespass. As noted in Chapter 8, the creation of the milk marketing boards, by providing a regular income and minimising the importance of location in relation to markets, further increased incentives to use pasture for dairy cattle, although some sheep were often retained in order to make the best use of grazing. Post-war, interest in sheep has somewhat revived on farms on the Downs and the Cotswolds, where greater acreages under leys have provided an incentive to keep them. In Gloucestershire, East Sussex and Wiltshire, for example, numbers approximately doubled in the 1950s and have risen further since; but densities are still low compared with upland counties, for a viable intensive sheep system that will enable sheep to compete on lowland farms has not yet been devised and accepted.

Only in the uplands, then, have sheep retained a prominent place in British farming and they are now more numerous than they have ever been, except in parts of Scotland, although much of the increase has taken place on the lower hills rather than on the open moorland and is associated with better grass management and heavier applications of fertilisers (Figure 10). The decline in parts of Scotland owes something to afforestation, which has been largely at the expense of rough grazing, but arises partly from a fall in the productivity of the upland grazings. There is, moreover, no necessary relationship between afforestation, which also provides shelter and may improve grazings, and the sheep population: sheep flocks in Argyll were maintained at about 700,000 between 1925 and 1955, despite the planting of over 75,000 acres with trees. Numbers in Wales have risen steadily since 1947, mainly on the hill margins, although a large acreage has been afforested.

What has changed in both uplands and lowlands is the composition of the sheep flocks; they now consist largely of breeding sheep and lambs, few of which see their first birthday. Sheep were originally kept primarily for their wool, but mutton had become the chief product by the mid-nineteenth century. Sheep flocks then included not only breeding sheep and lambs, but also many older wethers. In 1893-95, the first period for which statistics are available, there were 36 per cent. breeding ewes, 23 per cent. sheep 1-year-old and over (mainly wethers) and 38 per cent. lambs, but by 1961-65, the proportions had changed to 40, 13 and 47 per cent. respectively; if allowance is made for those sheep of 1-year-old and over intended as replacements, the number of wethers is now probably only a quarter as high as in the 1890s and most of these will be slaughtered as wether hoggs before they reach 15 months and not at 3 or 4 years as formerly. Since both ewes and lambs are more demanding and selective feeders than wethers, this change may have contributed to the deterioration believed to have occurred in the quality of upland grazing.

Sheep are unique in that sheep enterprises of a kind can be undertaken anywhere within Great Britain, but in present circumstances sheep cannot compete as a major enterprise with dairying on grassland or with cash crops on arable land, and will not be able to do so unless stocking rates and lambing percentages can be improved; the gross margin per acre for sheep on the lowlands (£12–£15 per acre) is only half that for barley (£27–£30) and a third of that from milk (£35–£40). Sheep may be retained on lowland farms because they fit into some ecological niche, e.g. steep or broken land which cannot easily be ploughed or used by dairy cattle. Alternatively, they may complement other enterprises to maximise grazing, to use surplus feed or to employ labour or land which would require large inputs of capital to be devoted to cereal-growing, for example, an additional combine. Similarly, while they compete with dairy stock for available grazing in spring, they can sometimes use winter grazing which heavier cattle will poach. They may also survive because farmers like keeping sheep or because it is traditional in that area; this is clearly a factor in the persistence of sheep in Kent, and G. Allanson has commented that 'many farmers will continue to keep sheep without continually counting their contribution to farm profits or the profits that might have resulted from some other enterprise'.

THE STRATIFICATION & SEASONALITY OF SHEEP FARMING

The place of sheep in the farm economy thus varies considerably in both importance and nature, for not all areas are equally suited to producing fat lambs. Within this general distribution of sheep there is some areal specialisation of function, related to environmental conditions, from which spring the movements which make it so desirable to consider the sheep economy of the country as a whole. As with beef cattle, there is a distinction between breeding and fattening, but since lambs may be slaughtered at any age between 2 and 15 months, there is generally no corresponding stage of purchasing sheep as store animals to be subsequently sold for fattening (though a survey in 1964 in Scotland, did find that some 10,000 lambs had been on more than two farms). Unlike beef production, however, there is a large-scale movement of breeding stock from upland to lowland.

There is thus a distinctive stratification of sheep farming, reflecting the relative importance of breeding and fattening and based on the characteristics of the different breeds and on the kind and quality of feed available, though it is not very amenable to statistical analysis. This stratification enables breeders in different areas to concentrate on only a few characteristics; it enables those uplands which cannot be used for other forms of agriculture to contribute to both the breeding stock and the supply of sheep for fattening; and it allows breed size and fertility to be matched with the environment and the feed available. The scale of the upland contribution to sheep farming in the lowlands in the early 1950s was estimated by B. R. Davidson and G. P. Wibberley at between $2\frac{1}{2}$ and $3\frac{1}{3}$ million sheep a year, of which just over a quarter were draft ewes and two-fifths store sheep; those draft ewes capable of further breeding contributed about half the lowland breeding flock at any one time. Such movements prolong the life of ewes of upland breeds, and provide meat better suited to market requirements than that of the larger breeds of the lowland.

The progression from upland to lowland is characterised by four features: the intermingling of breeds by crossing; lower losses and more lambs per ewe; a greater emphasis on fattening; and a declining emphasis on sheep in the economy of individual farms. There are exceptions, such as Romney Marsh, and great variety in any area, for much depends on the breeds in any locality and on the character of the environment, particularly the quality and availability of feed crops; wherever possible, fat lambs, which alone attract guaranteed prices, are produced. The following analysis is inevitably highly generalised; to give any adequate account, even if information were readily available, would require a large volume.

The first stage is the upland pastures, carrying pure-bred flocks of hardy mountain breeds. Records of hill sheep subsidy indicate that ewes of such breeds number nearly 5 million and contribute just over a third of the total breeding flock, the proportion ranging from over half in Scotland to under a sixth in England. Only these breeds can use such an environment, although there is a considerable choice of breed and a changing balance between them; for example, the Swaledale has gradually come to dominate much of the Pennines. Such flocks are self-supporting; ewes characteristically produce 3–4 crops of lambs before being drafted to the lowlands for fattening or for further breeding, although mortality rates are high. Lambing percentages are low, so that replacements for the flock have first claim on the ewe lambs. Formerly these were often wintered away on lowland pastures, although practices varied from region to region, depending on the quality of grazing, the proximity of low ground and the relative importance of rough grazing and improved land; for example, ewe lambs in the Pennines have generally been wintered at home and those in the Highlands sent to farms on the east coast. Such winter grazing is, however, increasingly expensive, and costs rose more than four-fold between 1939 and the mid-1950s; it is also becoming more difficult to find as stocking rates for cattle on lowland pastures increase, and home-wintering is undertaken wherever possible. Surplus ewe lambs and ewes which have borne several crops of lambs are drafted to lower farms with more improved land. Depending on the price of store lambs, wether lambs are fattened if possible, and the proportion similarly varies inversely with the harshness of the environment; where such lambs cannot be fattened, or where store prices are favourable, they are sold as stores in the autumn.

Conditions and practices vary from one upland to another. In the Highlands, where farms are large and have little improved land, the environment is too poor for sheep to winter on the hills and the farm is run as a single unit. In the Southern Uplands, better conditions permit more home-wintering and it is common for farms to be divided into hirsels, each with its own flock. In northern England, farms are again run as a single unit, but there is generally more improved land and ewes move to the better grazing for lambing. In Wales, where farms are often small, there is characteristically a third category of enclosed rough land, or *fridd*, between the improved land and the open moorland.

The lower hill country represents the second stage, although it is absent where transition from upland to lowland is abrupt. The $2\frac{1}{2}$ million ewes which are either returned as crosses or which qualify for hill sheep subsidy at the reduced rate in England and Wales give some indication of the size of the breeding flock here. Draft ewes and surplus ewe lambs from the mountain flocks are crossed with rams of long-woolled breeds, such as the Border Leicester or the Teeswater, to produce a first cross, which is larger and grows more quickly than the mountain breed, and so takes advantage of the better environment; where possible, these lambs are also fattened, but if the grazing is poor and insufficient feed is available, these too will be sent to lowland farms. The Border Leicester is the most important ram for crossing purposes and the Scottish Half-bred, the Greyface and the Welsh Half-bred are the first crosses from matings with Cheviot, Blackface and Welsh Mountain ewes respectively; such crosses accounted for a quarter of all fat lambs and hoggs sold in Scotland in 1965–66. In other areas, such as the Welsh Borderland, flocks of Clun, Kerry Hill and other pure-bred ewes are kept. On all these farms, sheep are only one of several enterprises and are generally subordinate to cattle.

On those few lowland farms where sheep are kept, the breed and importance of sheep varies from farm to farm and from area to area. It is on such farms, mainly close to the hill country, that most lambs drafted from hill and mountain flocks are fattened and half-bred ewes and ewe lambs are crossed again, often with a ram of a Down breed, such as the Suffolk, although the choice of breed depends on the first cross and on the farmer's preference. There are also pure-bred, self-replacing flocks of lowland breeds, where surplus lambs and ewes are finished on the farms on which they were born, as well as specialist ram-breeding flocks. Systems are very varied, for besides farms which support a regular breeding flock, some only fatten purchased stores and others winter ewe lambs. There are also flying flocks on farms with surplus feed in spring and autumn, where ewes, often drafted from mountain flocks, are purchased in the autumn and sold with their lambs in the spring. Types of fattening also vary considerably, although comprehensive data are lacking. Most lambs are probably fattened on grass, with supplementary feed, but 45 per cent. of those fattened in Scotland were folded on green fodder or roots, the latter almost entirely between December and February.

A distinctive feature of sheep farming (as of beef cattle) is thus the movement of livestock from one farm to another. Such farms may be at different elevations in the same valley and are often in the same ownership or occupation, the shepherd occupying the farmstead on the higher 'led' farm; but they may also be widely separated and sheep are sent from the great sheep fairs like those at Craven Arms in Shropshire or Hawick to farms all over the country. For example, in 1968, about a third of the store sheep sold in Scotland for fattening went to English farms, some as far away as Yorkshire, the EastMidlands and even East Anglia, and some 500–600,000 store sheep are said to be transferred from Scotland. Sheep from Wales similarly move to lowland farms in Wales, to the Welsh borderland and further east. In Scotland there are also large internal movements and a third of the lambs and hoggs sold in Scotland in 1965–66 were purchased as stores, a proportion rising in East-central Scotland to over half, mainly from Blackface flocks in South-west Scotland and the Highlands. Apart from migration within farms in early summer from low ground to high, and the return of ewe lambs wintered elsewhere, these movements are one way, predominantly easterly.

Romney Marsh, besides having the highest density of sheep, is also interesting because it presents the highland/lowland relationship, not only in miniature, but in reverse. It acts as a reservoir of breeding stock from which lambs are sent elsewhere for fattening or for wintering; for the Marsh is cold and exposed in winter. But here the 'upland' is the neighbouring parts of Kent, Sussex and Surrey and, unlike the practice in the true uplands further west, many sheep return to the Marsh the following spring for fattening.

An interesting aspect of sheep movements is the role of the market, although local differences in demand seem to exercise little direct influence. Paradoxically, sheep are relatively most important in Scotland, where little lamb and mutton is eaten (3·4 lbs. a head compared with 6·3 lbs. in Great Britain as a whole and 8·6 lbs. in London). The role of livestock markets was considered in Chapter 6, but even their location seems to exert little influence; for example, over a tenth of farms in a sample survey in Scotland were more than 50 miles from the nearest store market, a remoteness from point of first sale much greater than for other classes of livestock.

Both the movement of sheep from one stratum to another and the sequence of lambing and fattening have marked seasonal characteristics which vary from one part of the country to another, but with the main events occuring progressively later from south to north and from lowland to upland. Except for ewes of the Dorset Horn breed, which lamb in autumn, ewes lamb in spring, but the date of lambing ranges from the end of December to May, depending on average weather conditions and on farmers' policies. For example, on moorland farms on Exmoor most lambs were born in March and April, but on better land proportions were higher in February and March; similarly on a sample of farms in Gloucestershire, Wiltshire and Somerset in 1958–59 lambing dates ranged from January–February for early fat lamb production to late March and April for store lambs. The sale of fat lambs is also seasonal. Early lamb fetches a higher price and numbers marketed rise steadily from the end of March. The first lambs marketed in Scotland are crosses with Down breeds, sold from lowland farms

TABLE 21

PERCENTAGE OF GROSS OUTPUT DUE TO SHEEP AND WOOL IN ENGLAND AND WALES, BY TYPE OF FARM, 1966

	Northern England	Central and Southern England	Eastern and South-east England	South-west England	Wales	England and Wales
Specialist dairying	4	1	1	1	2	2
Mainly dairying	9	3	1	5	9	5
Livestock—mainly sheep	45	—	—	—	52	49
Livestock—sheep and cattle	29	21	9	31	33	26
Cropping—mainly cereals	7	6	4	5	16	5
General cropping	3	3	2	—	—	—
Mixed	6	5	6	12	19	7

Source: Farm Management Survey

in April, while the first Blackface lambs from hill flocks are sold in August. The seasonality of production and the movement of store and breeding stock are dictated largely by the availability of feed, especially on mountain pastures, which cannot support as many sheep in winter as in summer.

The importance of sheep and their relationship with other enterprises varies throughout the country and from one type of farm to another. In 1962, sheep and wool accounted for 15 per cent. of gross agricultural output in Scotland but 11 per cent. in eastern Scotland and 41 per cent. in the Highlands; for 1966 the corresponding proportion was 11 per cent. for Scotland, but values ranged from 93 per cent. on hill sheep farms, 43 per cent. on upland farms, 15 per cent. on rearing with arable farms, to 9 per cent. on cropping farms and 5 per cent. on dairy farms. Comparable data are not available for England and Wales, but Table 21 shows the proportions by major regions; these figures demonstrate the minor importance of sheep on most types of farming in the English lowlands.

TABLE 22
PERCENTAGE OF HOLDINGS RETURNING BREEDING EWES, BY TYPE OF FARM, 1966

ENGLAND & WALES

All full-time	Livestock	Mixed	Dairying	Cropping	Pigs and Poultry	Horticulture
40	90	62	33	24	15	15

SCOTLAND

All full-time	Hill Sheep	Upland	Rearing with Arable	Cropping	Dairying	Intensive
54	100	82	66	45	41	8

Source: Agricultural Departments

Similar conclusions emerge from an examination of the distribution of sheep on holdings of various types, confirming the general importance of sheep in Scotland, although the types of farming recognised officially in Scotland and in England and Wales differ (Table 22).

A more detailed regional analysis of England and Wales shows how the type of farm with which sheep are associated varies from region to region (Table 23). The figures should be read together with the percentage of the breeding flock found in each region and the proportion of full-time holdings having breeding sheep.

Sheep can be kept anywhere in Great Britain, but because of their minor role in the farm economy, the geography of sheep farming is exceedingly diverse. At the same time, it is highly regionalised and developments in one area can have a profound impact on others; the concentration on crop production in eastern England and the increasing stocking of cattle on lowland

TABLE 23
PERCENTAGE OF BREEDING SHEEP IN EACH REGION OF ENGLAND AND IN WALES, BY TYPE OF FARM, 1966

	England and Wales	Eastern	South-east	East Mid-lands	West Mid-lands	South-west	Yorks and Lancs	Northern	Wales
Livestock mainly sheep	28	+	22	11	10	11	23	26	49
Livestock sheep and cattle	33	10	14	25	40	37	17	43	36
Mixed	13	20	21	16	18	26	+	10	+
Dairying	14	+	+	13	17	17	33	15	+
Cropping	+	57	23	31	+	+	14	+	+
Percentage of breeding sheep in England and Wales	100	1	7	7	10	15	9	19	32
Percentage of full-time holdings with breeding sheep	40	5	24	38	40	40	41	69	78

Source: Ministry of Agriculture
+ =less than 10 per cent.

pastures have affected the demand for store sheep from the uplands, and a severe winter such as that of 1946–47 can greatly reduce the supply of sheep to the lowlands. Sheep play a major role only where no other enterprise is possible; elsewhere their role is minor and varied, and is dictated primarily by the characteristics of other enterprises.

FURTHER READING

E. A. Attwood and H. G. Evans, *Economics of Hill Farming* (1961)

B. R. Davidson and G. P. Wibberley, *The Agricultural Significance of the Hills*, Studies in Rural Land Use No. 3 (1956), Wye College

J. M. Dunn, 'Lamb and hogg production in Scotland', *Scot. Agric. Econ.* 17 (1967)

J. F. Hart, 'The changing distribution of sheep in Britain', *Econ. Geogr.* 32 (1956)

Natural Resources (Technical) Committee *The Sheep Industry in Great Britain* (1958)

CHAPTER 11

Pigs and Poultry

If sheep farming is the most extensive form of livestock farming in Great Britain, requiring much land and providing a low output per acre, keeping of pigs and poultry is the most intensive. Yet, while pigs and poultry account for a quarter of farm sales and a third of those of livestock and livestock products, they are rarely mentioned in geographical accounts of agriculture. This is understandable, for they are little affected by the environmental factors which are important in the distribution of most other enterprises. Moreover, pigs and poultry have traditionally been minor enterprises in various systems of mixed farming and since 1945, they have increasingly become self-contained activities having little connection with other enterprises; they have thus rarely dictated farming systems and mapping, not surprisingly, reveals no clear patterns in their distributions. These, however, are not haphazard and the increasing economic importance of pigs and poultry is sufficient reason for their separate treatment in this book.

Pigs and poultry have several similar characteristics, although their resemblance should not be pressed too far. First, since they do not depend on grazing, their direct requirements of land are small and they do not compete for land with other enterprises (although they compete for other resources). Formerly they depended largely on unsaleable, surplus or by-products of other agricultural enterprises, such as tail corn, and on domestic waste; but increasingly they have become little more than processors of purchased feed, much of it imported, although cereals, milled on the farm, have become more important in recent years, so that their indirect requirements of farm land are increasing, notably in East Anglia.

Their second common characteristic is the speed with which numbers fluctuate. Each year sows can produce two litters of pigs, averaging eight piglets, which reach maturity in nine months, although they may be slaughtered at any age from four months; this potential rate of reproduction compares with an annual output of one calf per cow and one or two lambs per ewe. For laying fowl, which average over 200 eggs per annum, the potential rate of change is even greater; the average life of broilers is only 60 days, and several batches can be produced each year. Numbers of both breeding sows and marketable pigs can rise sharply and, conversely, the slaughter of breeding stock and replacements can lead to rapid reduction of the national herd: for example, the number of pigs in Great Britain fell by three-fifths between 1938 and 1943 (3·8 to 1·6 million) and increased fourfold between 1947 and 1954 (1·3 to 5·4 million). This possibility of rapid variation underlies 'the pig cycle', which appears to be levelling out: high prices lead farmers to increase production, demand is satisfied but supplies cannot immediately be curtailed and prices fall, leading to a reduction in the breeding herd and a rise in prices. The fluctuations in the national poultry flock have been proportionally less marked, but the numbers of fowl nearly halved between 1938 and 1943 (60 to 32 million) and more than doubled again by 1951 to reach 73 million.

A third common feature is that the organisation of these enterprises has changed rapidly in recent years. In the past, pigs and poultry were kept in small numbers, often on small holdings, where they helped to increase gross output from a small acreage. Indeed, large numbers were kept on holdings of less than one acre, in back gardens and in other places where they escape enumeration in the agricultural censuses: probably about a fifth of the national flock of laying birds and perhaps a tenth of the pigs are not included in the agricultural returns, although these proportions are declining. Others are kept under fairly extensive systems on mixed farms, often acting as scavengers by eating what would otherwise be wasted. Most of these flocks and herds were small, but a rising proportion of both pigs and poultry is now kept in large units in factory-like conditions, industrialisation having probably gone further here than in any other branch of livestock farming. This trend has been strongest in poultry keeping and it becomes increasingly difficult to distinguish between the characteristics of large-scale poultry farming and those of manufacturing industries: for example, one projected broiler unit included 78 broiler houses, each 100 yards long, and investment was estimated at £1·7 million. Accompanying these developments have been a rapid rise in numbers of pigs and poultry and a fall in the numbers of holdings with them. Between 1951 and 1965 fowls increased from 73 million to 102 million and pigs from 2·5 million to 6·7 million, both the 1965 totals being well above pre-war averages. The number of holdings with pigs in England and Wales fell from 149,000 in 1955 to 82,000 in 1966 and those with poultry from 305,000 to 187,000 between 1952 and 1963. The changes in both pig and poultry keeping have led to reductions in costs and enabled pig and poultry meat to compete successfully with other meats; it has also permitted producers to satisfy the demand for processed and packaged foods, itself a consequence of improvements in living standards, changing working habits and the appearance of the supermarket. Before the Second World War, poultry meat was a luxury, but it is now one of the cheapest forms of meat, and per capita consumption has risen from 5·0 lbs. per annum pre-war to 16·7 lbs. in 1965–66. Consumption of pork has similarly increased from 12·3 lbs. in 1934–38 to 25·8 lbs. in 1965–66. Despite the rise in population since the 1930s, imports have diminished and British farmers now provide virtually all the eggs the home market requires, compared with only 71 per cent. of a smaller per capita consumption pre-war.

Pigs

The pig is not only very productive, but is both an efficient converter of vegetable matter into human food and omnivorous, a characteristic which allowed it to become established in a variety of areas where different kinds of feed were available. It provides the raw material for three kinds of products: fresh pig meat (pork), cured pig meat (bacon and ham) and manufactured pig meat, such as sausage and pies. Keeping bacon pigs is both the most distinctive and the most exacting branch, demanding expert management, skilled labour and careful feeding; but it is also the most static, for home production has increased by only a quarter since 1938 and per capita consumption not at all, whereas consumption of both pork and manufactured pig meat has risen sharply. A further distinction is that whereas home production of pork and manufactured pig meat is virtually sufficient, some two-thirds of bacon and ham requirements are imported, mainly from Denmark. Little pork is produced in Scotland, but available statistics are inadequate for detailed geographical analysis of these different branches, and pigs will be treated as a single enterprise. This procedure has some justification, for producers enjoy some flexibility in their choice of outlet and many bacon pigs are now used in part for manufacture.

THE DISTRIBUTION OF PIGS

A map of the distribution of pigs reveals no clear regional pattern (Figure 15). Numbers are small in and around the uplands of the west and north, and the highest densities are in parts of Yorkshire and Lancashire, the West Midlands, East Anglia and Greater London; densities are generally lower in northern England and Scotland, and the concentration in eastern and southern counties has become more marked in recent years. The actual distribution is more uneven than this map suggests, for pigs are kept in a minority of holdings, 27 per cent. in England, 22 per cent. in Wales and 11 per cent. in Scotland, with the highest value (33 per cent.) in Yorkshire and Lancashire and the lowest (1 per cent.) in the Highlands.

This distribution map can be further qualified by considering the relationship of pigs to other livestock, using livestock units (Figure 15); this comparison reveals that pigs are relatively most important in eastern England, especially East Anglia, and that south-west Cornwall is the only western area with high values. Examination of the number of pigs in herds of various sizes similarly reveals considerable regional differences and shows that, while herd sizes vary widely in all areas, there are proportionately more large herds in Eastern England and these account for a larger share of the pig population than further west; such herds are often found on large cereal farms. Herds of 200 or more pigs account for 22 per cent. of herds and, more significantly, for 72 per cent. of pigs in the Eastern Region, 7 per cent. and 49 per cent. respectively in South-west England, but only 2 per cent. and 27 per cent. respectively in Wales

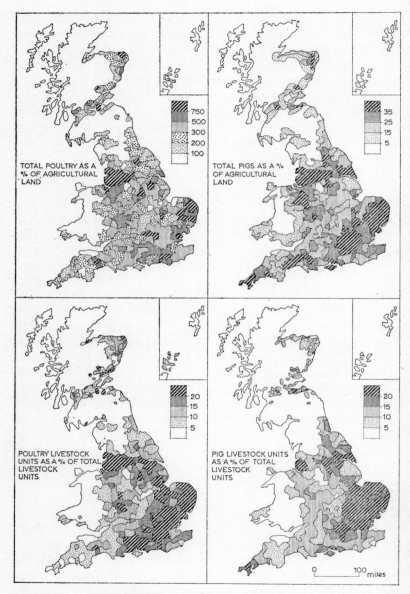

Figure 15. Pigs and poultry
Source: agricultural censuses 1965

(Table 24). A similar, though less marked, contrast is observable in Scotland.

TABLE 24
PIGS IN HERDS OF DIFFERENT SIZES IN 1967 AND 1968

ENGLAND & WALES (1968)

Herd size (pigs)	Eastern	South-eastern	East Midlands	West Midlands	South-western	Northern	Yorks and Lancs	Wales
			Percentage of holdings with herds of different sizes					
Under 50	44·6	45·7	63·6	68·4	69·2	70·4	61·2	88·4
50–99	17·5	16·5	15·1	14·2	15·1	14·0	16·3	7·3
100–199	15·9	15·6	10·9	9·0	8·6	8·2	11·1	2·8
200 and over	22·0	22·2	10·4	8·4	7·1	7·4	11·3	1·6
ALL	100·0	100·0	100·0	100·0	100·0	100·0	100·0	100·0
Herd Size			Percentage of pigs in herds of different sizes					
Under 50	5·4	5·1	11·4	15·0	17·0	16·4	11·2	40·5
50–99	8·1	7·5	12·4	14·5	16·0	14·9	12·4	18·0
100–199	14·8	14·1	17·8	18·1	18·2	17·2	16·4	13·8
200 and over	71·7	73·3	58·4	52·4	48·8	51·5	60·0	26·7
ALL	100·0	100·0	100·0	100·0	100·0	100·0	100·0	100·0
			Percentage of holdings with pigs					
	28	20	25	26	32	23	33	22

SCOTLAND (1967)

Herd Size	Highland	North-east	East-central	South-east	South-west
		Percentage of holdings with herds of different sizes			
Under 50	79·1	74·7	64·2	62·4	72·7
50–250	15·9	19·7	27·3	26·2	20·7
250 and over	5·0	5·8	8·5	11·5	6·6
ALL	100·0	100·0	100·0	100·0	100·0
Herd Size		Percentage of pigs in herds of different sizes			
Under 50	16·0	16·3	9·9	7·8	13·0
50–250	33·4	28·6	34·4	29·3	30·4
250 and over	50·6	55·0	55·6	62·9	55·4
ALL	100·0	100·0	100·0	100·0	100·0
		Percentage of holdings with pigs			
	1	21	16	18	13

Source: Agricultural Departments

Unlike sheep farming and, to a lesser extent, other enterprises based on grazing, the keeping of pigs shows no clear regional specialisation of function or seasonality of production, although both were probably more marked in the past, when considerable seasonal differences existed in the availability of feed, as with the production of milk in summer for cheese making, and fewer pigs were kept in protected environments; in the 1920s, producers of grain, potatoes and milk regularly bought store pigs for fattening. There is some specialisation of function between rearers and fatteners; there are also pedigree herds supplying gilts and boars for breeding, and increasing numbers of hybrid stock are being purchased by large national companies. According to a special analysis of the agricultural returns for England and Wales in 1954, no pigs were being fattened on a fifth of the holdings with breeding stock, and a sample survey by the Pig Industry Development Authority (P.I.D.A.) in 1960 showed that a third of the herds sampled depended to some extent on purchased weaners. Qualitative evidence also suggests that owners of small herds are more likely to concentrate on breeding weaners for sale and those of large herds on fattening, a view supported by the P.I.D.A. survey, which found that 36 per cent. of the herds specialised in breeding and 16 per cent. in fattening, and that breeding only was more common among the smaller holdings and fattening only among the larger, although both occurred in nearly all size groups. Such specialisation between holdings often occurs in the same locality, but differences in holding size and in the availability of feed also lead to some regional contrasts; high land values around towns may also discourage pig breeding. Unfortunately, there is no comprehensive information about the scale of inter-regional movements or the distances travelled, but there are well-known examples, such as the considerable eastward movement of store pigs from Cornwall.

FACTORS AFFECTING THE DISTRIBUTION OF PIGS

The reasons for these patterns of distribution are complex, reflecting not merely the interplay of contemporary forces but also the evolution of pig farming over the past century. This predominantly lowland distribution suggests that physical conditions may have some influence, but this is increasingly less true as more pigs come to be kept under wholly or partly controlled environments. Pigs are more sensitive to cold, damp and exposure than sheep or cattle, especially when they are very young; they use a higher proportion of their food intake to maintain body temperature in adverse conditions, an important fact when feed represents between two-thirds and four-fifths of total costs. These considerations help to explain the virtual absence of pigs from the uplands and their lesser importance in northern areas; a pre-war study in Wales, for example, found that, while both breeding and rearing occurred below 500', only rearing took place between 500' and 1,000', and few pigs were found above 1,000'. But most pigs are now kept in purpose-built (and sometimes heated) buildings; the housing of pigs for fattening is a long-established practice, and is increasingly important for other classes of pig. The P.I.D.A. sample survey found that, even in summer, 88 per cent. of herds were kept wholly indoors during farrowing and 69 per cent. during rearing, with still higher proportions in winter, although artificial heat was provided for only 55 per cent. of herds. Such housing represents a heavy investment and poses problems of disease control, and there is debate about the merits of extensive systems, which are still practised, especially for rearing, and are less demanding of labour. Where pigs are kept out of doors, the nature of the soil may be important; for example, pigs cannot so easily damage dry chalk soils as heavy clays.

Feed is by far the most important cost in commercial pig production, but the pull of sources of supply has weakened as feeding has become more specialised; diets of all classes of pig are becoming more sophisticated as increasing attention is paid to grading. The availability of feed, whether from farm or non-farm sources, was formerly considered important in locating the production of pigs, for it often consisted of waste or surplus products, such as sub-standard potatoes and skimmed milk, which were both difficult and costly to transport; a major non-farm source was domestic swill from urban areas, which had the merit of being regularly available. The relationship, however, was not simple and there were obvious anomalies, such as the comparative unimportance of pig-keeping on the peat Fens, where large quantities of potatoes were grown.

Such connections with crop surpluses are now less important, though there is some evidence that the recent expansion of pig-keeping in eastern England is associated with the alternative outlet it provides for barley on large arable farms; milling home-grown feed may yield savings of up to £4 per ton over the cost of purchased feed. Pig keeping was also associated with the making of butter, cheese and clotted cream on farms in Cheshire, Cornwall and elsewhere, although there has never been the close connection between dairying and pigs that has characterised Danish farming. Farmhouse manufacture of milk virtually disappeared after the creation of the milk marketing boards in 1933, and owing to the priority given to the liquid milk market and the consequent uncertain supply of manufacturing milk, no comparable association developed between factory and farm; pig keeping is not now closely linked with dairying. Nevertheless, pigs remain an important enterprise in many areas in which such relationships originated.

How far the location of markets exerts any influence is uncertain, although the pull of urban centres is undoubtedly less powerful than in the nineteenth century. Since much pig meat is either cured or manufactured, the immediate markets are often factories, whose location may originally have been related to the supply of pigs. A further complication is that links between both producers, whether in co-operatives or other groups, and between producers and processors of pig meat are growing closer. Large firms increasingly dominate the bacon and manufacturing trade: the 38 largest bacon factories produce over 80 per cent. of the bacon, and most bacon pigs are bought on contract. Groups of farmers are also combining to form larger selling units and weaner groups account for 12 per cent. of all weaners sold in Wales. How far such links have affected the pattern of movement of pig meat to markets is not known, but the practice of paying farmers a uniform price irrespective of location, adopted by F.M.C. Ltd., which has 40 per cent. of the bacon curing capacity and allocates pigs to factories, has led to long-distance movements and helped to continue production in remoter areas. The final market also reflects regional differences in consumption of pigmeat, with Scotland and Wales well below average and London and the Midlands well above, and such differences in demand have presumably exercised some influence on the location of production and may partially explain the limited development of pig keeping in Wales and Scotland.

A third significant factor is size of holding, though this, too, is becoming more complex; for there is no necessary connection between holding size and herd size. Pigs have been associated with small holdings since the Industrial Revolution, when commercial pig keeping began to emerge to meet the demands of rapidly-growing urban markets and pigs were commonly kept on small holdings near the main urban centres, and in small numbers on mixed farms. There has thus long been a dichotomy between size of holding and size of enterprise, but the association of pigs with small holdings was still marked when the Natural Resources (Technical) Committee made its report on scale of enterprise: 20 per cent. of the pigs in England and Wales in 1954–55 were on holdings with under 20 acres of crops and grass, although these accounted for under 5 per cent. of the acreage of all holdings, and 53 per cent. were on holdings of under 100 acres, accounting for 31 per cent. of the acreage. Holdings of under 20 acres still have 21 per cent. of all pigs, although the proportion varies from 28 per cent. in Yorkshire and Lancashire to 17 per cent. in Wales; but the share of holdings under 100 acres has fallen to 49 per cent while that of holdings of 300 acres and over has risen from 17 to 22 per cent.

The relationship between pigs and holdings has been complicated by the trends towards larger herds and holdings, which are encouraged by both technical and economic considerations. Because pigs require little land, large herds may be found on small holdings and vice versa; 27 per cent. of all pigs on holdings of under 20 acres in England and Wales were in herds of 500 or more in 1968, while a third of those in herds of 1,000 or more were on holdings of over 500 acres. Smaller holdings are thus often substantial businesses, and 39 per cent. of the breeding pigs in England in 1966 were on holdings requiring 1,200 or more s.m.d.s., compared with only 16 per cent. on those with under 275, the minimum required for a full-time holding. Pressures on farmers' profit margins in the past decade have made it necessary to improve efficiency, and the size of a viable pig-unit has risen from some 20–25 sows per man to 40–50, and farmers have either expanded the size of their herds or abandoned pig farming altogether. The adoption of industrialised methods has facilitated enlargement but required great capital investment; for an intensive pig unit may involve building costs of some £10,000. The trend towards larger but fewer herds, (Table 25) must be seen in the context of a rising pig population. No comparable regional figures are available, but the increasing proportion of the pig population in eastern and south-east England, which have over half the herds with 500 or more pigs, suggests that these trends are strongest there.

TABLE 25
CHANGES IN HERD SIZE IN ENGLAND AND WALES, 1955–66

Herd Size	Percentage of Herds		Percentage of Pigs	
	1955	1966	1955	1966
1–4	24·1	17·3	1·8	0·5
5–9	16·0	9·9	3·6	1·0
10–29	33·0	28·5	18·0	7·0
30–49	11·2	12·3	13·5	6·6
50–99	9·3	14·2	19·5	13·8
100–199	4·4	9·3	19·1	17·9
200–499	1·7	6·4	16·5	26·5
500–999	0·3	1·6	5·5	14·9
1,000 & over	0·0	0·5	2·5	11·8
ALL	100·0	100·0	100·0	100·0

Source: Ministry of Agriculture

As the scale of investment and the size of a minimum economic unit increases, it becomes more difficult both to begin and to give up commercial pig keeping; yet nearly half the herds are below this minimum economic size and a contributory factor is the varied combinations of resources on holdings of different sizes. The large holding will probably have a commercial pig unit employing at least one man full-time, but pigs on small holdings may utilise either labour which is not fully employed or unpaid family labour, and such production is less sensitive to price changes. This factor may also have a regional component: even in the 1920s commercial herds were more common in eastern counties, and in 1968 nearly half the breeding pigs in Eastern England and over half of those in South-eastern counties were in herds of 50 or more breeding pigs, compared with 12 per cent. in Wales and 26 per cent. in South-west England.

Clearly the relationships between pigs and types of farming are complex, and pigs are associated with many types. In 1954, 55 per cent. of the pigs in England and Wales were on holdings with three or four enterprises and 9 per cent. on specialist holdings with no other enterprise. In 1966 no separate category of specialist pig farms was recognised and pig and poultry farms accounted for under 4 per cent. of the full-time holdings (but 28 per cent. of the pigs); such specialists are most common in Yorkshire and Lancashire and in Eastern and South-east England. Understandably, the association between pigs and type of farming varies from region to region; in Eastern England cropping farms contain the largest number of pigs, in the West Midlands, South-west England and Wales dairy farms are the leading type and in the East Midlands, South-east England, Yorkshire and Lancashire, pig and poultry holdings. Formerly many farmers found pig-keeping a convenient, though not essential, enterprise; it is now becoming either a specialist activity or one of several major enterprises. Accompanying this greater concentration of pig production into few hands is an increasing geographical concentration in eastern counties.

The context of pig keeping has thus changed considerably over the past century and each change has contributed something to the present pattern. It is not the purpose of this chapter to consider the organisational changes that have taken place, but the efforts made to regulate production and to improve the lot of those farmers who keep pigs may be mentioned. The attempts in the 1930s to help the bacon industry by marketing boards were not very successful and were not repeated after the Second World War; but guaranteed prices for fat pigs were introduced after the war, and the Fatstock Marketing Corporation was formed and acquired an important place in the disposal of bacon pigs, P.I.D.A. was established and later the Meat and Livestock Commission, which absorbed it. Yet, despite success in increasing consumption of pig meat as a whole, two-thirds of bacon and ham is still imported, mainly from Denmark. The Danish industry enjoys no natural advantages over its British counterpart, although its close links with the dairy industry, particularly the return of skim milk from the creameries, provide the basis for a cheap and satisfactory diet for pigs; rather its advantages are organisational, for it is geared to the needs of its markets and is highly efficient, enjoying lower costs than British producers. Its success is a standing reminder of the importance of efficient marketing.

Poultry

The trends already noted in pig-keeping are much more marked in poultry farming, where industrialisation has gone further than in any other branch of agriculture. Nevertheless, like pigs, poultry are still kept in very varying conditions and, because of different attitudes and costs, small flocks persist, even though the back-yard producer continues to decline in importance as standards of living rise and recreational habits change. And, while the forces making for the concentration of production in large factory-like units are fairly well understood, it is no easier to find a satisfactory explanation for the present distribution of poultry keeping, which reflects the legacy of the past, wide differences in efficiency and the lack of clear locational advantages of particular parts of the country.

The term poultry includes ducks, geese, fowls and turkeys, but 95 per cent. of the birds enumerated in the agricultural census are fowls, the majority of them kept to lay eggs for human consumption. Less than 0·2 per cent. of the national poultry flock consists of geese and only 1·0 per cent. of ducks, though the latter also exhibit the trend towards the large-scale intensive methods of the broiler industry. Turkeys, accounting for 3·4 per cent of the national flock, are handicapped by their identification with the Christmas market, although their appeal is being widened and the trend towards large-scale intensive methods is quite marked, with nearly two-thirds of turkeys in flocks of 10,000 or more birds. Geese, ducks and turkeys, however, receive only passing mention in this section, which is devoted mainly to considering fowls.

The keeping of fowl can be divided into three branches: breeding, egg-laying and meat production, each having rather different geographical characteristics. Before the First World War, when the national flock was much smaller, most birds were kept in dual-purpose flocks, but specialisation is now highly developed. In recent years breeding has become concentrated in the hands of a small number of producers, of whom the more important operate on a large scale; less than 200 holdings have three-fifths of the breeding birds and 15 companies control most of the output. This branch exhibits no marked geographical concentration and, although important in the structure of the industry and responsible for much of its increasing productivity, will not be considered further. Keeping hens for egg production is the largest branch in both number of producers and output, contributing nearly two-thirds of sales of poultry and poultry products; but the balance between egg and meat production is shifting as consumption of poultry meat rises at the rate of about 10 per cent. per annum and will continue to grow, whereas the egg market is largely satisfied and demand inelastic.

Egg and meat production differ in several respects, but before they are considered separately, some general aspects of the distribution of poultry (which is virtually identical with that of fowls) will be examined. Poultry, like pigs, are widespread throughout the English lowlands (Figure 15), and there are no seasonal differences in distribution. Lancashire is the leading county and, with Cheshire, one of the main areas of production. Numbers are also high in many districts in eastern and southern England, and a belt of counties with much lower densities (which has its counterpart in pig distribution) extends from the Severn to the Wash and separates these two areas. Poultry are comparatively few in northern England, Wales and Scotland, although Orkney is an interesting exception, and this southerly distribution is becoming more marked. As with pigs, poultry's share of livestock units is highest in eastern arable counties, where there is comparatively little grazing. Such areas also have many of the very large flocks.

Considerable numbers of poultry are not enumerated in the agricultural census and a complete distribution map would give greater prominence to the urban areas. Furthermore, the tendency for poultry to be increasingly concentrated in large units means that their distribution has become more 'spotty' and is poorly represented by a choropleth map of average densities.

Reasons for this distribution, which is the resultant of several different components, will appear in the subsequent discussion, but some preliminary comments may be made. Most poultry are now kept in controlled environments, so that climate and topography have little direct effect upon their distribution. Even in the past, when poultry were kept mainly on free range, natural factors were not of major importance, although the greater severity of climate in northern and upland counties may have been a deterrent, for more food was required for maintenance and fewer eggs produced per bird. It is said that the inter-war rise of poultry keeping in the Fylde owed something to the flatness of its terrain, facilitating the use of folds and Lancashire cabins, and the erection of modern poultry housing is more difficult where there is little level land. But, in general, the correspondence between low densities and western and northern counties is related more to markets and other considerations than to physical suitability for poultry keeping.

Like pig-keeping, poultry farming has also been associated in the past with small holdings; in 1954, 32 per cent. of adult fowl in England and Wales were on holdings of under 20 acres of crops and grass, and 68 per cent. on holdings of under 100 acres. A similar relationship exists today, but there is no necessary connection between holding size and flock size, and large poultry units, many of which are substantial businesses, can often be found on quite small acreages. There is thus no clear spatial correspondence between poultry-keeping and the proportion of farmland occupied by small holdings, but the little land required by poultry and the low proportion of costs represented by rent do enable poultry keepers to incur high land costs in location on the outskirts of large urban markets; for proximity to markets has been a major influence in the past.

Although feed is the largest item of costs in poultry keeping, accounting for between 50 and 70 per cent., there is no clear relationship between the distribution of poultry and sources of feeding-stuffs. This is not now surprising given the dependence of many flocks on carefully prepared rations of purchased compounds; but even in the past such relationships were less clear-cut than for pigs. There was some association with cereal growing, where tail corn was available, and this probably helped in the more rapid increase in poultry farming between the wars in the main cereal-growing counties of eastern England. A similar relationship probably existed in the parallel post-war expansion of large-scale cereal-growing and poultry-keeping, for considerable economies can be effected in the home mixing of feed; but a more important reason for the expansion in eastern and southern counties was probably the existence of farmers with capital and business acumen who were not wedded to traditional methods.

Fowls, whether for egg or broiler production, are associated much more closely with specialist holdings than are pigs, although such holdings are few. Those classified as predominantly poultry in England and Wales account for only 2 per cent. of the full-time holdings, and most of them are in Eastern and South-east England (37 per cent.) and Yorkshire and Lancashire (19 per cent.). Concentration is less marked for fowls kept for egg-laying, where 42 per cent. of the birds on full-time holdings are on predominantly poultry holdings and 16 per cent. on pig and poultry holdings; concentration on specialist holdings is greatest in Eastern and South-east England and the East Midlands, with half or more of the fowls kept for egg-laying on predominantly poultry holdings, although only in Wales, where there are more fowl on dairy farms (29 per cent. compared with 24 per cent.), are such holdings not the leading type. In broiler production, 72 per cent. of birds on full-time holdings are on predominantly poultry holdings, and the proportion is again highest in Eastern and South-east England (85 per cent. and 84 per cent. respectively), exceeding 50 per cent. everywhere except the West Midlands (46 per cent.).

EGG PRODUCTION

The distribution of hens and pullets laying eggs for eating is widespread throughout the lowlands; Lancashire with 10 per cent. of the national flock, though little more than 1 per cent. of the agricultural land, is specially prominent. Forty-seven per cent. of the holdings in England and Wales in 1966 had a flock of fowl for the production of eating eggs, while in Scotland, despite the much smaller number of birds and the lesser importance of poultry farming, the proportion was 55 per cent. What these facts do not indicate is the prominence of large producers. While nearly 70 per cent. of the flocks in England and Wales and 75 per cent. of those in Scotland had under 100 birds, half the birds were in flocks of 5,000 and over, and in 1968, some 200 holdings of 20,000 or more birds accounted for a fifth of the birds kept for egg production.

The present geography of egg production reflects partly developments before the Second World War and partly organisational changes since the 1950s. Before the First World War egg production was both less important and less specialised, most flocks providing both eggs and table birds, and the scale of operation was small. No reliable statistics exist for this period, but the distribution of flocks seems to have been more uniform, although Lancashire was already the leading county. A possible explanation of this incipient localisation was the more dispersed settlement and industry in the valleys of the southern Pennines and a tradition of keeping small stock among industrial workers in north England, but this would not explain why Lancashire should stand out; its proximity to Liverpool, through which point-of-lay pullets from Ireland entered the country, was possibly a contributory reason. Whatever the explanation, the attractions of Lancashire were confirmed between 1914 and 1930. Before the First World War most eggs were imported, but the wartime blockade led to a rapid rise in the price of eggs, which encouraged many small holders to expand their flocks. Stimulated by high prices and by the industrial depression in the immediate post-war period, many such small holders, often former industrial workers, became specialist egg-producers. Further, the emergence of some pioneer poultry breeders in the western part of the county encouraged concentration on egg production in Lancashire and their success led to Lancashire's being accepted as the centre of poultry breeding, to the migration there of those wishing to specialise in the industry and to the adoption of egg production by many general farmers in that locality.

In the 1920s and early 1930s egg production expanded rapidly, encouraged by low prices for feeding stuffs and by the unprofitability of other enterprises, and imports were reduced. The number of fowl doubled between 1924 and 1934, with the fastest rate of growth in eastern counties and the slowest in Wales and Scotland; nevertheless, despite this differential growth, Lancashire remained by far the most important county. Most flocks were still small and capital investment in housing was low: more than 80 per cent. of the eggs marketed in the 1920s came from general farmers and small holders. The poultry flock was generally a minor enterprise, often looked after by the farmer's wife, for whom it provided pocket money. Production was highly seasonal for, in the free-range conditions then general, egg production fell off markedly in winter when climate was adverse and days short; for example, in 1935 production in the highest month was estimated at $2\frac{1}{2}$ times that in the lowest. Prices varied accordingly and only the specialist concentrated on winter production.

During the Second World War, numbers of fowl were sharply reduced to save imports of feeding stuffs, and birds not on farms increased to an estimated one-third of laying fowl. Immediately after the war, with egg-production profitable, the national flock grew so rapidly that there was over-production by the early 1950s, and this, together with the re-introduction of seasonal prices (favouring the specialist) and a growing awareness of new methods, led to marked organisational changes in the industry. With a guaranteed market for all eggs, falling prices encouraged farmers to increase the size of flocks to maintain incomes. Concurrently, intensive methods enabled those who could take advantage of them through good management and capital investment to reduce costs per bird, and encouraged such producers to enlarge their flocks. The processes of enlargement and intensification were further stimulated by suppliers of young chicks and of feeding stuffs, who either entered a kind of share-cropping arrangement in which they owned the flocks or provided loans and other financial inducements.

Although intensive methods were known before the Second World War, when some half million birds were in battery cages, most flocks in the late 1940s were still small and 95 per cent. were kept on free range. Almost 90 per cent. of all holdings in England and Wales had poultry, and 80 per cent. of these flocks had under 100 adult birds. Many producers turned to cage batteries or deep litter houses which, by providing a controlled environment, led to more eggs per bird, especially during winter, when prices were higher. They also led to reduced labour costs, although capital costs were increased; fixed equipment for a flock of 20,000 birds represents an investment of £25,000. By 1967 only 6 per cent. of birds were kept on free range, 21 per cent. were in deep litter houses and 73 per cent. were in cage batteries. This change in method, together with improvements in feeding and in the quality of stock (80 per cent. of which are now hybrids), has led to a sharp increase in the productivity of birds, from 172 eggs per hen in 1957–8 to 204 in 1966–7.

TABLE 26
SIZE OF FLOCKS OF ADULT FOWL IN ENGLAND AND WALES

		Flock Size		
	1–99	100–499	500–	All
		Percentage of holdings		
(a) Adult fowl				
1948	85·5	13·6	0·9	100·0
1957	63·4	13·6	3·1	100·0
1966	68·9	22·5	8·6	100·0
		Percentage of birds		
1948	46·6	42·2	11·2	100·0
1957	20·3	50·9	28·8	100·0
1966	7·8	16·7	75·5	100·0
(b) Fowls producing eggs for eating		Percentage of holdings		
1964	61·8	28·9	9·3	100·0
1968	71·1	19·4	9·5	100·0
		Percentage of birds		
1964	8·4	24·3	67·3	100·0
1968	5·7	10·8	84·5	100·0

Source: Ministry of Agriculture

Such intensive methods demanded a large flock to spread overheads and enlargement of flocks was made feasible by the abolition of the rationing of feeding stuffs in 1954. The proportion of birds in large flocks consequently increased very rapidly (Table 26). Most of the smallest flocks are still kept on open range, and the average size of flocks on free range is only a fortieth of that of flocks kept in cage batteries.

The figures in Table 26 are not strictly comparable because of changes in definition, and do not reveal the contribution of very large flocks. This increase in size has led to a considerable reduction in holdings with laying fowl, particularly among the smaller flocks; the total number in England and Wales fell from over 280,000 in 1958 to some 150,000 in 1966, though the national flock increased by one-third in that period.

This concentration on fewer holdings has been accompanied by a geographical concentration. The number of fowl in laying flocks has fallen in strictly rural areas like Cornwall, Cumberland, rural Wales and even East Anglia, and is increasingly concentrated in the axial belt linking Greater London with Lancashire; for these larger units are often located near the main urban markets.

Although the profitability of flocks of different sizes and kept under different systems varies widely, with as great a range within groups as between them, and with management the most important variable, there are nevertheless advantages in large size, and 30,000 birds probably represent the optimum, with few economies of scale for larger flocks. Large producers obtain larger discounts on purchases of day-old chicks and of feed (some 2s. 6d. per cwt. in one study), they make better use of labour and they receive bonuses from packing stations for their larger deliveries. Against this, the small producer may enjoy low costs for feed, housing and labour if he can use existing buildings, tail corn and unpaid family labour. Given the great variety of holdings, costs must vary widely; for what is suitable for one producer with surplus labour or a shortage of land may be quite inappropriate for another.

The role of the location of the market in egg production is both complex and interesting (Chapter 6). Since eggs are fragile and need to be marketed as fresh as possible, one might expect markets to exercise a major influence on the location of production. This was certainly true when flocks were small, the market poorly organised, transport relatively slow and costly, and direct sales to consumers and retailers the major outlet; even so, the relationship was not simple and a map showing the distribution of poultry in 1887 (which has many statistical imperfections) does not reveal the presence of either Newcastle or Glasgow. More recently the marketing of eggs has been complicated by contractual relationships between buyers and producers, whether by contracts to packing stations or by co-operative ownership of packing stations; some third of the eggs marketed through packing stations pass through producer-owned co-operatives. A further complication is seasonal differences in demand; for example, Cornwall is an importer of eggs in summer, because of the holiday trade, and an exporter in winter. More important have been the arrangements for implementing guaranteed prices for eggs, which have led to a marked distinction between the 60 per cent. of eggs marketed through the Egg Marketing Board and the 40 per cent. sold otherwise. The arrangements by which eggs are sold to the Board permit production in areas where it would not be possible in a free market (Chapter 7). By contrast, those producing the 40 per cent. of eggs which do not use the Board's machinery are near, or at least have easy access to, large urban centres.

It will be interesting to see what effect the abolition of the Egg Marketing Board and the subsidy on eggs has on the geography of egg-production; almost certainly, in the absence of a guaranteed price and the Board's marketing arrangements, production will be concentrated into fewer hands and be much more market-oriented than at present. Such concentration will create considerable social problems; for example, the greatest dependence on egg-production is in Orkney, the most remote of all the producing areas, where 96 per cent. of flocks have fewer than 500 layers, but where eggs account for a fifth of agricultural output.

MEAT PRODUCTION

The rise of the broiler industry is a striking feature of post-war British farming. As already noted, poultry production before the First World War was unspecialised, but the production of poultry meat subsequently became quite subordinate to that of eggs. There were specialist producers of table birds, but much of the poultry meat came from culls and old hens from the egg-laying flocks, and per capita consumption of poultry meat was low. Specialists virtually disappeared during the Second World War and post-war expansion was restricted by the rationing of feeding stuffs. Then, from 1954 came the rapid emergence of the broiler industry. This had its origin in 1947 in a report by an official mission to the United States, where there had been a rapid increase in the 1940s in the production of young chickens killed at 9–14 weeks and a decline in the output of costly roasting chickens. This change had been stimulated by an appreciation of the rapid rate of growth of fowl during their first ten weeks and by advances in poultry feeding. The end of rationing brought a similar rapid expansion in Great Britain. In 1953–4 some 5 million broilers were produced, compared with 14 million other table birds. By 1959–60 the comparable figures were 100 million and 8 million, and the output of broilers now exceeds 200 million.

The broiler industry is both more localised than egg production and concentrated into fewer hands. In 1968 broilers were found on less than 2 per cent. of holdings in England and Wales and 3 per cent. of those in Scotland. The great majority are very large businesses and it is believed that one producer is responsible for 15 per cent. of the output. Fewer than 400 producers produce over 80 per cent., and 55 per cent. of the broilers in England and Wales and 59 per cent. of the much smaller number in Scotland are in flocks of 50,000 birds and over. The area of many of the holdings on which these flocks occur is small; two thirds of the flocks of 50,000 and over in England and Wales are on holdings with under 50 acres of crops and grass.

Land requirements are thus low and labour accounts for only 4 per cent. of total costs, for one man can look after 20,000 birds. By contrast, the capital cost of buildings is high, e.g. £12,500 a unit for 20,000 birds. Many who took up broiler production were businessmen who saw in it a profitable investment and much of the growth of the industry has thus taken place outside the traditional poultry areas, mainly in east and south-east England and often within easy reach of large urban markets (although Norfolk and Huntingdon are among the leading counties). Few broilers are produced in western and northern areas, where farmers lack resources and are remote from processing plants. This growth of the broiler industry is primarily responsible for the south-eastward shift of the centre of gravity of poultry production as a whole.

The broiler industry is characterised not only by the large size of its units but also by the closeness of its links to markets. Most of the output is purchased by a few buyers controlling retail chains and, since most birds are sold ready for the oven, they have to be processed in large factories which require a uniform product and level supplies; contracts between producer and buyer are consequently a characteristic feature. Other producers have formed groups to own processing factories and these groups sometimes prescribe a minimum output for membership. In other instances buyers have acquired their own production units.

This great expansion in the production of poultry meat is unusual, not only for its rapid growth and for the large scale and the industrialised methods adopted, but also because it has occurred without government support. Unlike red meat, poultry meat does not enjoy a guaranteed price, though it receives some protection from the regulations governing the import of foreign supplies, nor are there any arrangements for controlling marketing, as with egg production.

CONCLUSION

These developments, which have made pigs and poultry the second most important sector in British agriculture, have also made geographical analysis, on the basis of available data, more difficult. This does not mean that they can safely be ignored by geographers, but the problems of location and linkages with other enterprises which they present now resemble those of light industry and require data of a quite different kind. If American experience is any guide, they may also indicate the direction in which other branches of livestock farming may go, although the latter are unlikely to achieve the close control over production in factory-like conditions exemplified by the modern egg battery or broiler house.

FURTHER READING

D. B. Bellis, 'Pig farming in the United Kingdom', *J. Roy. Agric. Soc.* 129 (1968)

J. C. Bowman, *The Egg Production Industry in the United Kingdom 1945–67*, Dept. of Agriculture, Univ. of Reading (1960), typescript

R. Coles, 'The poultry industry today', *Assoc. Agric. J.* 8 (1966)

Ibid, *Development of the Poultry Industry in England and Wales 1945–1959* (1960)

Report of the Reorganisation Committee for Eggs, Cmd. 3669 (1968)

CHAPTER 12

Crops

Crops, other than horticultural produce, provide 18 per cent. of farm sales in the United Kingdom, though only 11 per cent. in Scotland. Cereals account for some 60 per cent. of these sales and cash roots for most of the remainder. These estimates understate the importance of crops, for large quantities are retained on the farms on which they are grown. Some 46 per cent. of the wheat, 76 per cent. of the barley, and 86 per cent. of the oats are used for animal feed, often on the farm of origin, and there are also some 1·2 million acres of other crops grown for stock feeding. Crops retained on the farm provide 23 per cent. of the concentrated feedingstuffs fed to livestock on farms; indeed, home-grown crops provide, through subsequent processing of crops sold, a further 30 per cent. of concentrates fed on British farms. This chapter, however, is concerned mainly with cash crops.

In acreage, crop production has fluctuated considerably in the long term, but, in general, these trends have been fairly steady, in spite of fluctuations from year to year, depending on the weather and other natural hazards such as the east coast floods in 1953. Until the Second World War, trends in crop yields were less marked, but the Second Agricultural Revolution, with the introduction of new varieties, new methods of disease, pest and weed control, higher applications of fertilisers and greater mechanisation of crop production, has since led to a sharp upward movement. Annual fluctuations are generally more marked in yields than in acreages, especially among root crops; for example, the average yield of sugar beet in Great Britain was 30 per cent. higher in 1960 than in 1959 and that of potatoes 25 per cent. higher in 1959 than in 1958; variations in any one locality can be very much greater. Such fluctuations can have serious repercussions, especially for potatoes, in which Great Britain is virtually self-sufficient.

Mono-culture or near mono-culture has rarely prevailed in Great Britain, so that two questions on the geography of crop production arise: why should varying proportions of land be devoted to annual crops, and what determines the choice of crops and the proportions in which they are grown? These questions are obviously related, for the expansion of the acreage under tillage crops since the mid-1950s has reflected the relative profitability of large-scale cereal growing. Each crop has its own requirements, but there are common factors, discussed in Chapter 7, which help to determine whether land is devoted to tillage crops.

On land economically feasible to plough, a further decision must be made about which crops to grow. In the nineteenth century, rotations were prescribed in farm leases, which specified severe penalties if they were disregarded, but now only activities which are against the rules of good husbandry and likely to harm the land can bring conflicts. In any case, half the land is now farmed by owners, who are free to cultivate the land as they wish. Rotations were developed to keep the land in good heart, both through restorative crops such as clover and by the provision of fodder for sheep and cattle, which returned manure to the soil. Rotations also helped to control weeds, crop pests and diseases and, by integrating crop production and livestock, spread demands for labour more evenly throughout the year. Fertilisers, herbicides and pesticides have made many of these functions unnecessary and crop land can be (and is increasingly) farmed without livestock. Rigid rotations have largely disappeared and the rapid rise in farm rents and land values has made farmers in the main arable areas anxious to achieve the maximum acreage under cash crops. Economic forces are also encouraging greater specialisation and fewer crops are grown per farm. Under the old rotations no field was ever planted with the same crop in successive years and any duplication of crops within the rotation was rare, but growing cereals year after year is increasingly common; more than twenty-five successive crops of barley have been reported and a survey in Bedfordshire in the 1960s found that 38 per cent. of farmers were taking at least four cereal crops in succession. Such practices have caused misgivings about possible effects on soil structure and plant hygiene, especially in areas where it is difficult to find alternative cash crops, as on the thin soils of the Hampshire and Wiltshire Downs. Leys provide the most obvious alternative, but the return from them is generally much lower than that from cash crops, especially if they are grazed by beef cattle or sheep.

If rotations no longer dominate choice of crops, other considerations play a part. Crops differ in their physical requirements, and these can be met with varying degrees of satisfaction in different localities, although most crops do well in the areas best suited to tillage, such as the Fenland. Average yields can indicate suitability, though yield is much affected by management and the administrative areas for which yield data are available are heterogeneous. A further complication is that where only a small acreage is grown, farmers may select fields and manage crops with greater care, so that yields may be higher than a broad view of environmental conditions would suggest. Quality, too, may not coincide with high yields, as with potatoes grown on the peat Fens, which, though high yielding, are not of the highest quality, or with sugar beet, where high yields generally mean low sugar content.

The effects of physical suitability are somewhat modified by comparative advantage, although this is difficult to determine with numerous crops, some of which are not sold. Thus yields of nearly all crops are high in the Holland Division of Lincolnshire, but the proportion of cereals is much lower than on the chalk downs, for vegetables and cash crops, which provide a higher return per acre, are preferred. Where crops are not sold, farmers may be influenced by a comparison of yields or of costs of production or, where these are not known, by labour requirements. Unfortunately, the only data available for an examination of the broad pattern of costs and returns relate to large regions, with too wide a range of conditions to provide a very meaningful analysis.

Marketing links may play a part, especially where crops are bulky or perishable or have limited outlets, as with sugar beet (Chapter 6); but the flexibility and cost of modern road transport and the methods of determining transport charges greatly weaken such links. Restraints on production under various marketing schemes, such as the acreage quotas allocated by the Potato Marketing Board and the Hop Marketing Board and the contracts awarded by the British Sugar Corporation, also affect the choice of crops.

Tradition and established practice may also be important, although their influence, too, is weakening as the number of farmers diminishes and those who remain become more cost-conscious. The continued, though declining, importance of mixed corn in south-west England is an example, for there is otherwise no satisfactory explanation of the popularity of this crop. Conversely, the unimportance of fodder beet and oil seed rape, both major crops in continental Europe, seems partly due to the fact that they are not traditional crops here.

CEREALS

Cereals are the leading cash crops, measured both by sales and by acreage, for they occupy three-quarters of the land under tillage crops in Great Britain and over 90 per cent. in Wiltshire. They have long accounted for most of the tillage, but these high proportions are of recent origin: 62 per cent. of the tillage acreage was in cereals in 1870 and only 59 per cent. in 1938. There have also been marked changes in the relative importance of the different cereals, notably a six-fold increase in the area under barley, which now comprises two-thirds of the cereal acreage, compared with a fifth in 1938, and is the leading cereal nearly everywhere. The competition among the three cereals illustrates clearly the complex economic, physical and technological factors determining comparative advantage.

The three main cereals (small quantities of mixed corn and rye for threshing are also grown) have many common features. They are easily mechanised and require little labour or attention except at sowing and harvesting; they are thus comparatively easy crops to grow and are not very exacting in their physical requirements, for cereals of one kind or another can be grown on most ploughable land. They are, therefore, relatively most important where soil conditions are difficult for other crops, as on thin chalk or stony soils, and least important where deep, easily worked and stone-free soils are pre-empted by more remunerative cash crops. Proportions are also low, however, towards the margins of cultivation, where the small acreage of tillage may be largely under forage crops. The main restraints are thus climatic rather than edaphic, arising chiefly from the decreasing prospect of successful harvesting as rainfall increases and the growing season shortens.

The expansion of cereal growing, however, largely reflects changes in the attitudes of farmers towards the place of cereals on arable land and in the balance of economic forces. Cereals were formerly important as a key element in most rotations, providing the main, sometimes the only, cash crop. Now they are increasingly grown because the economies of scale that arise from mechanised production offer one way of combating rising costs. With the widespread adoption of combines and their progressive replacement by larger and faster machines, labour requirements of cereals have fallen both absolutely and relatively, from about a third of total costs in 1947–8 to about a fifth in 1965–6. The contribution of rent, on the other hand, has doubled in the same period from about a tenth to a fifth, primarily through the sharp rise in rents since 1957. Machinery is now the largest element in costs of cereal production, at almost a third, having doubled since 1947; and once expensive combines and driers have been bought, it pays to use them to capacity to spread overheads over the largest possible acreage. The average acreage of cereals per grower has risen steadily and increased by two-thirds between 1960 and 1967; at the same time, the number of growers of cereals has declined.

Wheat, the bread grain, was long the most important cereal and the keystone of the economy of most arable farms, a position it lost to oats in the 1880s and to barley in the 1950s. Wheat is primarily a cash crop, for some 90 per cent. is sold and deficiency payments are made only on sales of millable wheat; it is used for both human food and animal feed in roughly equal proportions. Approximately half the total consumption is imported, since British wheat is generally unsuitable for bread making and is used only in limited quantities as a filler owing to its low gluten content. Some two-fifths of the British wheat used for human consumption is employed in breadmaking, but this provides only 20 per cent. of the flour used; a third of the wheat is employed for cake and biscuit making, which depends mainly on the home-grown product, and the rest is consumed as flour in the home. Home-grown wheat also provides nearly half the wheat used for animal

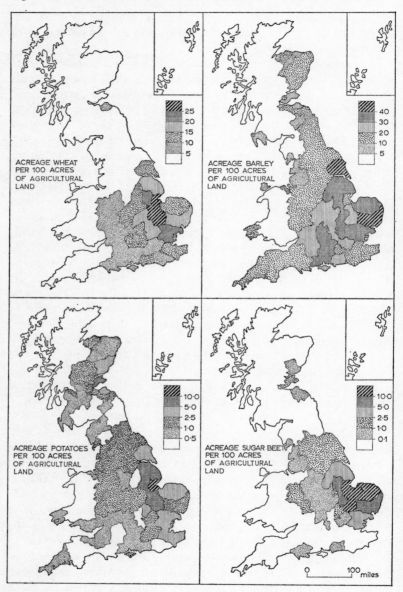

Figure 16. Crop acreages
Source: agricultural censuses 1965

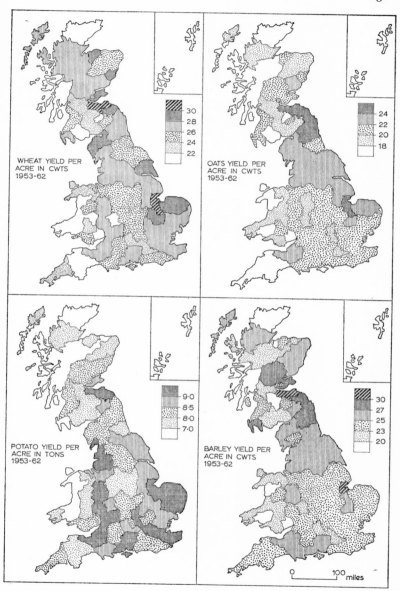

Figure 17. Crop yields
Source: agricultural censuses 1965

feedingstuffs. Although greater care in marketing and developments in baking technology might increase the share of home-grown wheat, there is clearly no prospect that British farming will be able to replace the wheat now imported.

Wheat is now grown mainly in eastern and southern England, especially the Fenland. It does not do well on acid soils and is susceptible to fungal diseases such as 'take all' and 'eyespot' if grown too frequently. Wheat is associated particularly with the better land and with those areas where, on average, summers are warmer and drier; the acreage in northern counties is limited by the amount that can be sown in autumn and by the fact that wheat is harvested later than barley. Within the areas where it is a major crop, wheat is relatively more important on the heavier soils, where its strong rooting system enables it to do better than other cereals. The advantage is relative rather than absolute. According to the National Wheat Survey drainage was the most important cause of variations in yields on farms growing wheat; on poorly drained soils yields averaged 30·7 cwt. per acre and on excessively drained 31·4, compared with 38·5 on moderately drained soils, while clays averaged 32·6, compared with 38·8 on fine loamy soils. Wheat is reputedly the most difficult cereal to grow, but winter wheat is the highest yielding cereal in all the main grain growing areas, attaining an average of 32·4 cwt. per acre in Great Britain in 1965, compared with 30·0 for barley and 24·6 for oats. By contrast, yields from spring wheat are much lower. Table 27, showing the proportion of holdings growing some wheat and the average acreage grown on full-time holdings in each region, emphasises the importance of eastern England, although data for Scotland show that the proportion of wheat growers in eastern Scotland is surprisingly high in view of the small percentage of the wheat acreage there. Most holdings with wheat in western and northern regions have only a small acreage; nevertheless, wheat is more strongly associated with large farms in all regions than is any other cereal.

Yields are a major variable in the profitability of cereal production: a poor crop costs almost as much to grow as a good one, and Figures 16, 17 and 18 emphasise this point and qualify the acreage maps. Unfortunately the data on which they are based, relating only to counties, have serious limitations. These are accentuated by the method of representation, which weights each county equally irrespective of the acreage grown, a fact of special significance in Scotland, where the information for large Highland counties such as Inverness and Ross relates mainly to a small acreage on the east coast. No clear pattern of wheat yields emerges, though, if appropriate allowance is made for these large Scottish counties, yields tend to be lower in the west and higher in the east. The range of variations is much less than on the acreage map, with a marked clustering around the mean, probably because management is a major variable and because the factors determining yields are very complex and vary throughout the country; for example, yields in south-eastern counties may be depressed by a deficit of soil moisture, and those in western and northern counties by excessive rainfall.

The National Wheat Survey, a sample survey of wheat growers in England, found little variation in costs, yields or returns in the different regions, with the exception of south-east England, where returns and yields were lower and part of the crop was grown on poor Weald Clay. This survey, however, cannot throw much light on why the proportions of wheat growers and of wheat acreage vary so much from region to region, for the probability of a grower being selected in the sample was proportional to his acreage and nearly half were in East Anglia. Respondents in western districts were probably limited to areas well suited to this crop, for most of the wheat grown elsewhere is in small acreages. Only 11 per cent. of occupiers in northern England grew wheat and 64 per cent. of this was grown on less than 50 acres per holding compared with 30 per cent. and 25 per cent. respectively in eastern England; most farmers in western and northern counties regard wheat growing as a risk not worth taking. Further, the data relate only to one year, when the weather at harvest time was unusually favourable.

Barley, by far the most important crop on British farms, is primarily a fodder crop and British farmers supply virtually all the country's requirements, apart from a specialised trade in malting barley, mainly from Canada; indeed, in recent years, a considerable, though variable, export trade has arisen. Barley does, however, compete with other cereals imported as feedingstuffs. The crop is used for both animal feed and human food in the ratio of about 5 to 1. Barley for human consumption is used mainly for malting and distilling and this demands a higher quality barley, which commands a premium and is generally achieved only with lower yields. Before the Second World War the main aim of barley growers was to obtain a malting sample, but this no longer holds and only a fifth of a sample of growers in 1967 grew primarily for quality, compared with two-thirds who grew primarily for high yields.

Barley is much more widely grown than wheat, but the main areas of production, whether by acreage or as a proportion of all cereals, are East Anglia and the chalk downs of Hampshire and Wiltshire (where the highest proportion of holdings with barley is found). Barley is intolerant of high soil acidity, the most common cause of crop failure. It is mainly spring sown and the proportion of spring barley increases northwards as the growing season shortens (Figure 18); for it is possible to sow late and still obtain a reasonable crop.

Oats, now the least important cereal and out-yielded by wheat and barley, are grown primarily as fodder, and only a small proportion leaves the farm; as with barley, price guarantees are implemented through acreage payments rather than on sales. The ratio of animal feed to human food from oats is about 11 to 1; the only important outlet for human consumption is for porridge oats and oatmeal.

The acreage under oats is more uniformly distributed than that of either the other cereals, but oats are relatively most important in western and northern districts, in some of which they are still the leading cereal. Oats are the least sensitive cereal, able to tolerate acid soils and low sunshine and capable of drying in the sheaf. Yet, as Figure 17 shows, oats do better in eastern counties, where they occupy little of the cereal acreage (although the range of county yields is even less than for wheat and barley). Oats are unlike the other cereals in that the proportion of growers is highest in western and northern areas; 45 per cent. of holdings in the Highlands have some oats, compared with 11 per cent. in eastern England, and the contrasts in the proportion of the acreage under oats falling into different size groups are also less marked. They are also more strongly associated with smaller holdings than the other

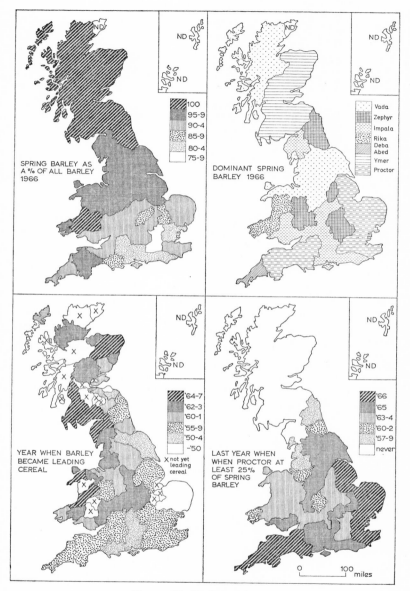

Figure 18. Barley varieties
Sources: agricultural censuses and Cereals Sales Censuses, Peter Darlington Partners Ltd., seed brokers. 'N.D.' signifies 'no data'

cereals. Oats are thus a residual legatee, largely grown where other cereals do not do well, although the exclusion of barley from the list of approved crops for winter-keep schemes has probably contributed to the continued importance of oats in upland districts.

THE CHANGING IMPORTANCE OF WHEAT, BARLEY & OATS

The importance of the different cereals, however, cannot be understood by considering them either in isolation or in a static context. Wheat was the leading cereal in the main arable areas in the mid-nineteenth century because it was the most remunerative cereal crop. It lost this position when imports of cheap high-gluten wheat from the New World flooded the British market and depressed prices, and it was widely replaced by oats, most of which was for consumption as fodder on the farm. There was then only a limited demand for barley for malting, but a malting sample could be obtained only in areas with relatively high amounts of sunshine, and barley was also not acceptable as feed for cattle. The expansion of the wheat acreage in the 1930s under the stimulus of the Wheat Act, which provided price support for a specified acreage of wheat, was largely at the expense of barley. At that time oats out-yielded barley and the weak-strawed varieties then available were liable to lodge unless the crop was grown on poor land; the low-base status of many soils after years of depression and neglect favoured the more tolerant oat crop and oats were reputedly more suitable than barley for harvesting by binder.

In the post-war period the relative advantages of the two cereals changed. New Scandinavian varieties of barley, with short, stiff straw, gradually replaced the native barleys. They often gave higher yields than oats and were less liable to lodge than the older varieties, and so were less affected by bad weather. They were well suited to harvesting by combine, which was then replacing the binder in the principal cereal growing areas. At the same time, traditional outlets for oats were declining and new demands were arising for barley. The replacement of horses by tractors led to a steady fall in the number of horses, for which oats were a major source of fodder, and better management of grassland weakened the demand for oat straw. Farmers gradually appreciated the higher yields and greater nutritional value of barley, and overcame their reluctance to feed barley to cattle, especially dairy cattle; the rapid expansion of the pig herd also led to a demand for barley. Oats received less attention from the plant breeders and some promising new varieties were late-ripening, a disadvantage in those northern districts where much of the oat crop is sown; yields of oats increased less rapidly than those of either wheat or barley. So, from a peak of 3·7 million acres in the Second World War, the oats acreage dropped steadily and in 1966 was only 23 per cent. of that in 1942, while the barley acreage expanded from 1954 onwards.

The increase in barley growing spread outwards from those counties in southern and eastern England where barley survived as a major crop and which are still the most important areas; Figure 18 shows this gradual northerly and westerly extension of barley by mapping the year when it became the leading cereal. This spread was helped by the great improvement in the base status of soils in the higher rainfall areas, resulting from heavy applications of subsidised lime during and since the Second World War, and by the increased number of combines in western and northern counties; by 1950 there was at least one combine for every thousand acres of cereals in all southern and eastern counties of England and by 1956 throughout England and Wales and in much of eastern Scotland. The delay in replacing oats by barley in northern counties partly reflects the previous importance there of oats and the extent of the change that was necessary before the acreage of barley exceeded that of oats. Oats survive as the principal cereal only in upland and northern districts, where the land is less suitable for combines, the small size of fields hinders their efficient use and farms are too small and have too little land under tillage to justify investment in a combine (though the role of the contractor in harvesting crops on small farms must not be overlooked); binders still outnumber combines in many western and northern districts and over 80 per cent. of the oat crop in Scotland was harvested by binder in 1964 (Figure 6). Although oats may fail in dry districts, they are still grown in eastern and southern counties largely because they are less susceptible than wheat and barley to fungal diseases and so are used as a break crop; the acreage under oats has accordingly made some recovery recently.

There has been a continual improvement in the varieties available, and the replacement of improved varieties by better ones can be mapped. Figure 18 shows the replacement by other varieties of Proctor, a most successful new barley, which reached its peak about 1960. Each cereal variety has somewhat different characteristics, which may make it more suitable in some areas than in others, and some idea of the complex pattern of preferences, derived from information supplied by Messrs. Peter Darlington Ltd., is shown in Figure 18.

The wheat acreage has changed little since the early 1950s, although wheat has been displaced by barley as the leading crop everywhere in the lowlands, except the Fens; it has maintained its position where it does well, and yields have risen almost as fast as those of barley. Yet, although the economic forces encouraging farmers to enlarge their cereal acreages apply to both wheat and barley and although winter wheat out-yields barley, wheat has not contributed to the rapid expansion in the cereal acreage. There are two principal reasons for this. First, outlets are more restricted for wheat than for barley. As has been noted, the amount of home-grown wheat used for human consumption is mainly limited by the inability of bakers to increase significantly the proportion of home-grown wheat in bread. Since wheat, unlike barley, must be sold to qualify for deficiency payments, wheat is virtually confined to being a cash crop; more would probably be fed to stock if the guarantee were paid by acreage, as with barley and oats. The method of paying the wheat subsidy, based on the difference between average selling price and guaranteed price, also means that the returns to individual farmers are directly affected by the yields they obtain, so encouraging wheat growing where it is favoured by soil and climate; with barley, where the same subsidy is paid whatever the yield, such considerations matter less.

A second major factor is the lesser susceptibility of barley to fungal diseases. Winter wheat cannot be cropped continuously without serious risk of loss from cereal diseases or a rapid increase in perennial grass weeds and many farmers still believe wheat should be grown only after a root crop or ley and followed by another non-cereal crop. As a result, wheat fits less easily into simplified farming systems and profits less from the economies resulting from the fullest use of capital invested in combines, grain driers and grain storage.

Barley was also favoured during the late 1950s by changes in the guaranteed prices for the two cereals. When feedingstuffs were derationed in 1954, the guaranteed price was 18 per cent. higher for wheat than for barley, but in 1957, in order to encourage home production of feedingstuffs, the price for barley was raised above that for wheat and remained higher until 1965, although the margin of advantage was gradually reduced. Nevertheless, the acreage continued to rise until 1967, for the technological factors encouraging the expansion of barley had apparently acquired a momentum of their own independent of price changes.

Guaranteed prices are now higher for both wheat and oats than for barley and there is increasing concern about the possible consequence of continuous barley growing. In 1967 the barley acreage fell for the first time since 1954 and the fall continued in 1968. This may be a temporary phenomenon, but it is salutary to note the responses of a sample of farmers, quoted in the Britton report, to changes in price guarantees. When the 1967 Price Review increased the price of wheat by 6d. a cwt. and reduced that of barley by 7d. a cwt. only 4 per cent. reported that the 1967 Price Review had had any effect on their cereal growing in 1967 and, although 28 per cent. of the others claimed that the Review appeared too late to affect their plans, the proportion whose plans for cereal growing in 1968 were affected was only 10 per cent., rising to 19 per cent. among those growing 300 acres and over. Twenty per cent. of those not intending to change explained that cereals were not the main business of their farm, but 19 per cent. claimed that nothing in the Review suggested a change of plan, 18 per cent. that they planned several years ahead, 13 per cent. that they were unwilling to chop and change from year to year, 11 per cent. that their plans were governed by their rotations and a further 7 per cent. that they were already growing what was best for the land they farmed.

The changes in the relative importance and distribution of the cereal crops illustrate the importance of approaching
s

questions of comparative advantage with caution. The reasons why one crop enjoys an advantage over another are complex and not always measurable in economic terms, while the response of farmers to economic changes is very variable. Nevertheless, the sequence of changes is sufficiently clear to show that farmers as a whole have acted consistently and intelligibly. They have found it worthwhile to grow more cereals and to expand their barley acreage, partly by substituting barley for oats and for non-cereal crops and partly by bringing more land under cultivation; in western counties, where the acreage under tillage crops has declined since the Second World War, the expansion of barley has been effected entirely by substitution for other crops and in eastern counties, where the proportion of land under tillage has risen since the mid-1950s, the increase in barley has been achieved both by substitution and by ploughing grassland. In these eastern counties there is now little unploughed land suitable for cereal growing. These advantages of cereals over other crops and of the various cereal crops over one another are not permanent; the distribution of cereals has been very different in the past, when much more land in western and northern counties was under cereals.

ROOTS

Apart from vegetables, which are considered in the chapter on horticulture (though they are extensively grown on crop farms), sugar beet and potatoes account for most of the remaining sales of crops, with 11 per cent. and 22 per cent. of crop sales respectively.

Although not technically roots, but tubers, potatoes share many characteristics with root crops and are generally considered with them. Root crops differ in several important respects from cereals. They are much bulkier, with yields generally ranging between five and ten times the weight per acre of cereal crop; their acreages are correspondingly smaller, totalling only 1·4 million acres (1·1 million for *cash* roots) compared with some 9·2 million acres under cereals. Transport costs are thus an important consideration and would be more so if producers paid the full costs of transporting roots from farm to market; as noted in Chapter 6, the transport policy of the British Sugar Corporation allows some sugar beet to travel quite long distances, although by far the greater part of the crop is grown within forty miles of factories, and the cost of potato haulage is much influenced by the possibility of return loads. Because the harvested part of the crop is grown within the soil, the nature of the soil is much more important than it is for cereals and restricts the growing of cash root crops to a limited range of soils. Traditionally, root crops have been cleaning crops because they require a great deal of cultivation, which helps to suppress weeds, but chemical sprays have made this much less important than formerly. Both for this reason and because of their bulk, they have also had heavy requirements of labour, although this need, too, is changing with the rapid advances in mechanisation, especially of harvesting.

SUGAR BEET

Sugar beet is important both as a cash crop, which provides almost a third of British requirements of sugar, and for animal fodder, both from the tops (which can either be fed or ploughed in as green manure) and from the dried pulp, which growers can buy from the factories in proportion to the beet they supply. The monopoly of the British Sugar Corporation as buyer of sugar beet for processing, the acreage limit imposed by the Government in the light of commitments to overseas suppliers, especially from the Commonwealth, and the embargo on the opening of new factories play an important part in determining where the crop is grown. This relationship will be strikingly demonstrated by the closure of the Cupar factory in 1971, for this will effectively eliminate beet growing in Scotland. The acreage now grown in England would undoubtedly be larger without such restraints, and the British Sugar Corporation estimates that there are at least a further 40,000 acres of suitable land in eastern England. A policy of spreading the benefit of growing sugar beet more widely, however, has led to a system of transport charges which enables the crop to be grown in areas where it would not otherwise be economically feasible, notably in southern England (Chapter 6).

Sugar beet is more localised than cereals, in both the areas and the holdings where it is grown (Figure 16). The crop is produced mainly in East Anglia and the Fenland, with a quarter of the acreage in Norfolk alone; there are also small outliers in the west Midlands and Fife, based on the factories at Allscott, Kidderminster and Cupar, and on the south coast. Sugar beet is grown on only 9 per cent. of the holdings in England and Wales and on less than 1 per cent. in Scotland and, as with other cash crops, the larger holdings have a disproportionate share. In England and Wales over a third of the crop is grown on holdings with 500 or more acres of crops and grass and 54 per cent. on holdings of 300 acres and over, compared with 16 per cent on those of under 100 acres; the proportion of holdings with beet increases with size of holding, reaching 38 per cent. on holdings of 1,000 acres and over. The distribution of the sugar beet acreage is similar, for holdings with under 10 acres of beet account for over half the holdings growing the crop, but provide only a tenth of the acreage, whereas the tenth with 50 or more acres of sugar beet provide two-fifths of the acreage. Moreover, production is being progressively concentrated into fewer hands; the number of growers fell by a third between 1956 and 1967 and the average acreage of beet rose from 11 acres to 18 acres.

Although government policy determines the acreage grown and the location of factories is influential in defining the areas where growing sugar beet is profitable, the physical requirements of the crop are also important. Unlike turnips, it does not do well in the moist and cool conditions of the west and north, where both yield and percentage of sugar content are low, important considerations when contract price rises with yield and sugar content. Yields of sugar beet delivered at the different factories (which draw most of their supplies from their immediate vicinity) are higher in eastern England than in the west Midlands and Scotland. Yields in Scotland are normally appreciably lower than in England, where summers are considerably warmer, and in 1964–66 averaged some 30 per cent. less. Sugar beet also does best on soils of high base status, in which a good seed bed can be prepared, which are deep and well-drained so that full root development is possible and which are free of stones, which deform roots and make harvesting difficult. The ideal is a deep calcareous loam, and in 1957 some 46 per cent. of the crop was grown on loams. A proportion is grown on fairly heavy soils, which may present considerable problems at harvest both because of the difficulties in using machinery, especially on the 10 per cent. grown on clays, and because of the amount of dirt which clings to the roots. In 1961, for example, dirt accounted for 23 per cent. of the weight of sugar beet delivered to the Felsted factory, which receives mainly beet grown on heavy loams in Essex, compared with 12 per cent. from Fenland beet. Some beet is grown on sands, where yields may decline sharply in dry years; chalk soils, although satisfactory in base-status and texture, are generally too shallow.

Labour costs represent a substantial part of total costs (over 30 per cent.) and account for 35 per cent. of variable costs; fertiliser, the second most important variable cost, accounts for 26 per cent. though only 11 per cent. of total costs. Labour costs are almost equally divided between casual and regular labour, with two major peaks of requirements, in spring, for hoeing and singling, and in autumn, for harvesting. The adoption of monogerm varieties, which do not require singling, reduce the former and mechanisation of harvesting the latter; 94 per cent. of the crop was mechanically harvested in 1966, compared with 57 per cent. in 1957. These improvements led to a reduction in the standard man-days per acre of sugar beet from 17 in 1958 to 10 in 1968.

POTATOES

A distinctive feature of sugar-beet growing has been the stability of acreage, which has not fallen below 400,000 or risen above 460,000 since 1942; by contrast, the acreage under potatoes, the other major cash root crop, has more than halved from a peak of 1·3 million acres in 1948. Perhaps the most important characteristic of potatoes is fluctuation in yield, of major significance in a crop which is mainly unprocessed and does not store well for more than the succeeding winter; for which demand is singularly unresponsive to price change, so that the same acreage can produce a glut or a deficit; and in which Great Britain is now mostly self-sufficient (except in early potatoes, about a third of which are imported before the home-grown crop is available). Controlling supply is a principal function of the Potato Marketing Board, which can specify the acreage to be grown, prescribe the minimum size of potato to be marketed and purchase surpluses for processing or stock feed (Chapter 6). The per capita demand for potatoes greatly increased during the Second World War, reversing a downward trend characteristic of countries with a rising standard of living, and is still higher than pre-war, although it will probably decline further.

There are in fact three potato crops, each differing somewhat in distribution and characteristics, seed potatoes, early potatoes and main-crop potatoes; but since seed potatoes (about a tenth of output) cannot readily be distinguished from either early or maincrop potatoes, a proportion of each being used for seed, they are shown together in Figure 16. Potatoes are widely grown (a reflection in part of the different requirements of the three categories and of the growing of small quantities for domestic consumption), but the main areas lie in the east of Great Britain; the Eastern Region in England contributes 28 per cent. and potatoes occupy over 10 per cent. of the agricultural land in the Holland Division of Lincolnshire. Measured by relative importance, however, Scotland (where potatoes are the leading cash crop) is much more prominent, as are the areas around the major cities, where potatoes occupy three-quarters of the land devoted to root crops. This wide dispersal, compared with sugar beet, is reflected in the higher proportion of holdings returning potatoes, which rises from 20 per cent. in Wales and 22 per cent. in England to 45 per cent. in Scotland and 59 per cent. in East Central Scotland. Data for England and Wales also show the greater importance of small holdings, for those with under 100 acres of crops and grass have 40 per cent. of the crop, and of small acreages of potatoes, holdings with under 10 acres of potatoes accounting for 18 per cent. of the acreage. The distribution of main crop potatoes is similar to that of all potatoes, but under 5 per cent. of holdings have early potatoes; of those in England and Wales two-thirds grow less than 4 acres, accounting for 11 per cent. of the acreage, and only 22 per cent. is grown on holdings with 50 or more acres. The proportion growing seed potatoes is probably even smaller and in Scotland, the main source of seed potatoes, the ratio of holdings on which seed potatoes are grown to those with ware potatoes only is approximately one to four.

Some rigidity is imposed upon the distribution of potato growing by the operations of the Potato Marketing Board (Chapter 6). These considerations matter more when output is expanding than when the acreage is declining, but they make it difficult for farmers who have not grown the crop in the past to enter production. The number of registered growers is declining and fell 37 per cent. between 1956 and 1966, so that, despite a declining acreage, the average acreage per grower rose from 8 acres to 11 acres and the proportion grown on holdings with 50 acres or more of potatoes is rising.

Yield is a most important factor determining the return from potato growing. Physical conditions are major causes of differences in yields and a Scottish survey of potato-growing in 1965 and 1966 shows a direct relationship between yields and the acreage grown. Although deep, well-drained soil is best for potatoes, they are less demanding than sugar beet in respect of soil conditions. They do not do well on thin or wet soils, and stones make harvesting difficult and deform the tubers; harvesting is also more difficult on heavy clays. On the Fen silts an exceptionally high proportion of the land in crops is under potatoes and yields are among the highest in the country; the limitation here is the risk of eelworm infestation, which increases greatly if potatoes are grown too frequently. The Fen peats are also high yielding, but the potatoes are of poor quality; by contrast, potatoes from the silt Fens are highly regarded and buyers on some markets will pay a premium for them.

Climatic differences are the principal reason for specialisation on early, seed or main crop potatoes. Early crops, although lower yielding, command a high price if sufficiently early, though it falls rapidly as the season advances; early potato growing is therefore something of a gamble. Areas of mild winters and early springs, such as the coasts of Ayrshire, Cornwall and Pembrokeshire, are therefore important and earlies account for over half the potato acreage in such areas where sites are selected for aspect and shelter rather than for ease of cultivation. However, the importance of these areas should not be exaggerated, for the bulk of the early crop is grown elsewhere, notably in Lincolnshire.

Potatoes are severely affected by virus diseases, most of which are transmitted by insect vectors, notably aphids. Consequently much of the acreage grown for seed (accounting for about a fifth of total cost and up to two-fifths of variable costs) is located in cooler areas where aphids are less likely, notably in Scotland. About a fifth of Scottish annual production and nearly half of its seed output goes to England and Wales, which obtain additional supplies from Northern Ireland. There is now, however, a wider interest in various upland areas in England and Wales and control of aphids by sprays is making the import of seed potatoes from other areas less necessary, a matter of some concern to Scottish farmers, who produce different varieties as seed potatoes for the English market from those which are sold as ware potatoes in Scotland.

In the principal areas producing main crop potatoes, the eastern lowlands of England and Scotland, the main climatic consideration affecting yield is probably water supply; potatoes are a crop it pays to irrigate in dry years. The risk of blight, a fungal disease that is the worst scourge of potato growers, is less in these areas, although all districts may suffer in wet seasons: the likelihood of attack can now be predicted from meteorological conditions and prophylactic spraying undertaken, and growers of large acreages of susceptible varieties find regular spraying justified.

The map of average yields of main crop potatoes (Figure 17) illustrates the range of differences, though no very clear pattern emerges. Yields are generally highest along the east coast and in a belt of country from Lancashire to Somerset, much of it in the rain-shadow of the Welsh uplands, where potatoes are largely grown on light textured soils; yields in western and northern districts are generally appreciably lower.

Labour is the biggest single item in the cost of producing potatoes, accounting for about 30 per cent. of total costs, almost equally divided between regular and casual labour; the graph of labour demands throughout the year shows one peak in spring, when potatoes are planted, and a much higher peak in autumn at harvest time. In one example, over two-fifths of the total labour requirements were used in October, for although great progress has been made with mechanisation, illustrated by the decline in the number of standard man-days for an acre of potatoes from 20 to 15 between 1958 and 1968, a large casual labour force is still necessary to pick and sort potatoes. Many of the former sources of casual labour have ceased and the presence of such a labour force, with a long tradition of casual work, is probably an important factor in the concentration of potato growing in the arable areas of eastern England.

The cost of transporting a bulky crop, largely sold for direct human consumption in the main urban areas, is predictably an important factor, and the relationship between large towns and the percentage of the root crops under potatoes has already been noted. The principal area of production is almost equidistant between the two major urban markets, Greater London and the conurbations of the Midlands and south Pennines (which rate more highly than their population suggests because of higher consumption per head); but the published evidence on the sources of supply is scanty and some potatoes move considerable distances. An increasing proportion is being manufactured as crisps or dehydrated potatoes or is being canned and, as with vegetables, the location of factories will be a major determinant of production for these markets; output of crisps, where there is a large reduction in weight and bulk, increased six-fold between 1955 and 1966 and now accounts for nearly a tenth of the potatoes used for human consumption.

As with the crops already considered in this chapter, there is no exact correspondence between production and the areas best suited to the crop; this is particularly noticeable near towns, where potatoes displace other root crops, and also in the balance between sugar beet, potatoes and vegetables in East Anglia, all high value crops which might each justifiably monopolise large tracts if physical suitability were the only criterion. Lacking satisfactory data, one must assume that each enjoys a comparative advantage over other crops, arising from the complex of factors that have been considered.

HOPS

Hops account for little over 20,000 acres, only 0·2 per cent. of the land under tillage crops, although they contribute about 2 per cent. of the income from crop sales. Production is much more localised than that of any other cash crop, with approximately two-thirds of the acreage in south-east England and one third in Herefordshire and Worcestershire. Their distribution, like their total acreage, has changed little for fifty years and largely reflects the availability of deep, well-drained soils and sources of manure and casual labour, which were formerly required in large quantities. These considerations are much less important now that the crop is largely harvested mechanically, but the pattern of production has been closely controlled since 1932 by the Hop Marketing Board and imports are also regulated. The hop marketing scheme is unusual in that each grower, who must be licensed with the Board, is given a basic quantity, not a basic acreage; this quantity attaches to the farm and is sold with it, but the actual proportion of this quota the Board will buy is prescribed from year to year.

OTHER CROPS

The remaining acreage under tillage is mostly under a fluctuating, though generally declining, acreage of bare fallow and various fodder crops. The latter, some 836,000 acres in 1966, now represent only 7 per cent. of the acreage under tillage crops, compared with 29 per cent. in 1875, when there were over 4 million acres, and 17 per cent. in 1938. They are relatively most important in western and northern counties, a very different distribution from that of a century ago, when these crops were mainly grown in the south and east, and include succulent root crops (turnips, swedes and mangolds), leafy brassicas (kale and rape), and leguminous crops (peas and beans), which provide protein. Each of these crops differs in its requirements and distribution; thus, turnips do well in the cooler conditions of Scotland and northern England, where they are still eaten off by sheep, while mangolds do better in the warmer sunnier conditions of southern England. Similarly, while rape is important as a pioneer crop and as sheep feed on the upland margins, kale is chiefly grown in southern England as feed for dairy cattle.

Their changing importance arises from various causes, including competition from imported feeding stuffs and from other home-grown crops, greater flexibility in the feeding of cereals to livestock, chemical weed control (reducing the need for cleaning crops), improved grass management and the search for a larger proportion of cash crops. Growing fodder root crops, which have not lent themselves so readily to mechanisation, has been particularly handicapped by high labour costs and only kale has increased its acreage substantially. The search for suitable break crops on cereal-growing farms has brought renewed interest in field beans, while the winter-keep scheme of 1964 has also increased the attractiveness of fodder crop production in northern and western counties.

FARM CROPS & TYPE OF FARM

As we shall see in Chapter 13, the distinction between horticultural crops and other crops grown for sale is somewhat arbitrary, but it is convenient to consider fruit and vegetables separately; the principal cash crops will therefore be regarded as wheat, barley, potatoes, sugar beet and hops. Figure 20 shows the proportion of all farm activity (measured in standard man-days) attributable to these cash crops; this distribution is more markedly easterly than that of tillage as a whole.

The crops mapped in Figure 20 are present in varying proportions in different areas and on farms of different types. Table 27 shows the place of wheat, barley and main-crop potatoes in England and Wales and in Scotland as a whole, and records, for each main type of farm, the percentage of holdings on which each crop is recorded, their share of the crop acreage and the average acreage grown. As would be expected, such crops are found mainly on mixed and cropping farms and these farms have the highest proportions of holdings with each crop and the largest average acreage; of the three crops, barley is the most widely dispersed among the different types of holding.

Table 28, which shows the percentage of holdings with wheat, barley and potatoes in Wales and the regions of England, should be studied in conjunction with Table 29, which gives for each region the percentage of the tillage acreage and of the acreages under each crop in England and Wales. It also, incidentally, indicates the relative competitive strength of the different crops, which would otherwise be present in the same proportions as the regional share of the acreage under tillage crops. For all three crops the highest percentages of holdings with each crop are on cropping farms, although the proportions are lower in the west and north than in the east. The proportions on holdings of other types are also highest in the eastern regions, so that wheat is grown on a higher proportion of dairy

TABLE 27
CASH CROPS AND FARM TYPE

ENGLAND & WALES, 1965

	Dairy	Livestock	Pigs and poultry	Cropping	Horticulture	Mixed	Part-time	All
			Percentage distribution of farm crops					
Wheat	14	6	2	55	4	16	3	100
Barley	17	7	3	48	2	18	5	100
Potatoes*	7	2	1	66	9	11	4	100
			Percentage of holdings with each crop					
Wheat	21	20	19	84	21	52	7	22
Barley	46	44	34	94	24	83	15	36
Potatoes*	22	24	10	59	33	45	9	21
			Average size of enterprise, in acres					
Wheat	25	28	28	58	27	36	8	36
Barley	26	30	37	86	27	50	11	42
Potatoes*	2	1	5	18	8	5	1	7

SCOTLAND, 1964

	Dairy	Upland	Rearing with arable	Cropping	Intensive	Arable rearing and feeding	Part-time	All
			Percentage distribution of farm crops					
Wheat	15	2†	9‡	63	2	8	1	100
Barley	20	3†	18‡	43	2	12	2	100
Potatoes	20	2†	10‡	53	2	8	5	100
			Percentage of holdings with each crop					
Potatoes	67	50	76	93	38	86	41	54
			Average size of enterprise, in acres					
Potatoes	6	2	3	22	6	6	1	5

* Main crop and second earlies only
† Hill sheep and upland
‡ Rearing with arable and rearing with intensive livestock

Source: Agricultural Departments

and livestock farms in the Eastern Region than of cropping farms in Northern England. On the other hand, except for wheat in the two regions of north England, the average acreage of each crop is larger on cropping farms in all regions than on dairy and livestock farms in any region, although these averages likewise decline from east to west and northwards. The proportion of the acreages under the different crops on cropping farms similarly declines, so that in eastern regions

TABLE 28
WHEAT, BARLEY AND POTATOES AND TYPE OF FARMING IN ENGLAND AND WALES, 1966

	Dairying	Livestock	Cropping	Horti-culture	Mixed	All full-time
			Percentage of holdings with wheat			
Eastern	51	48	90	30	79	66
South-east	36	46	80	11	72	39
E. Midlands	33	53	84	24	75	55
W. Midlands	23	44	70	16	64	33
South-west	16	16	61	5	30	19
Yorks & Lancs	7	11	71	14	48	26
Northern	8	10	46	2	28	16
Wales	4	6	22	5	15	6
			Percentage of holdings with barley			
Eastern	79	61	93	32	93	76
South-east	58	62	92	18	92	54
E. Midlands	61	72	97	31	95	76
W. Midlands	51	65	94	22	88	57
South-west	45	67	95	13	86	53
Yorks & Lancs	24	24	98	27	81	58
Northern	49	46	99	19	86	58
Wales	39	26	82	12	61	36
			Percentage of holdings with main-crop potatoes			
Eastern	10	3	54	42	32	42
South-east	10	6	29	21	24	15
E. Midlands	21	7	62	39	33	34
W. Midlands	16	19	62	24	43	24
South-west	16	28	25	15	43	22
Yorks & Lancs	12	7	71	30	45	28
Northern	31	21	72	40	53	36
Wales	31	28	42	31	55	22

Source: Ministry of Agriculture

most of the acreage is on cropping farms, while most of the much smaller acreages in other regions is on dairy, livestock and mixed farms.

Finally, an alternative interpretation of the place of these

TABLE 29
TILLAGE, WHEAT, BARLEY, POTATOES AND SUGAR BEET IN ENGLAND AND WALES, 1966

Percentage of acreage

	Eastern	South-east	East Midlands	West Midlands	South-west	Yorks and Lancs	Northern	Wales	England and Wales
Tillage	29	14	16	9	12	7	9	3	100
Wheat	35	16	20	10	9	6	2	1	100
Barley	25	15	16	9	14	10	9	3	100
Potatoes	36	7	19	10	4	14	8	2	100
Sugar Beet	66	—	19	6	—	6	3	—	100

Source: Ministry of Agriculture

TABLE 30
SALES OF FARM CROPS, 1964

ENGLAND & WALES

	Dairying	Dairying (Wales)	Predominantly Livestock	Mixed	Arable	Fen and Warp Arable	Limestone Arable
			Percentage of gross output				
Farm crops	8	3	8	36	68	74	75
			Percentage of output from crops				
Wheat	32	16	17	29	24	22	20
Barley	41	29	25	38	30	10	46
Potatoes	15	32	33	15	20	33	15
Sugar Beet	2	—	5	8	19	25	16
ALL CROPS	100	100	100	100	100	100	100

SCOTLAND

	Dairying	Hill Sheep	Upland	Rearing with arable	Rearing with intensive livestock	Arable Rearing and feeding	Cropping
			Percentage of gross output				
Farm crops	11	1	7	20	22	35	58
			Percentage of output from crops				
Wheat	8	—	—	3	12	8	15
Barley	36	—	—	37	47	41	29
Potatoes	26	2	15	16	15	23	32
Sugar Beet	—	—	—	—	—	—	6
ALL CROPS	100	100	100	100	100	100	100

Source: Farm Incomes England and Wales and Scottish Agricultural Economics

crops on farms of different types and of their competitive strength is provided by Table 30, showing the proportion of gross output (farm sales and deficiency payments) derived from each crop on the principal types of farms; this information, derived from the Farm Management Survey, relates to a sample of just over three thousand holdings. In England and Wales the proportions of gross output from crops on the generalised farm types range from 8 per cent. to 68 per cent. and in Scotland from 1 per cent. to 58 per cent., reflecting in part the less favourable environment for crop growth in Scotland and the greater emphasis on livestock; for individual types in England and Wales, the range is from 3 per cent. on Welsh dairy farms to 75 per cent. on light land arable farms on the limestone uplands of Lincolnshire. The table also shows the unimportance of sugar beet on all types of farms in Scotland and the minor place of wheat, even on Scottish cropping farms, where potatoes are the principal cash crop. Other interesting contrasts are in the relative importance of wheat and barley on arable farms on Fen and Lincolnshire warp and on the Lincolnshire limestone uplands respectively and the predominance of root crops over cereals on the former. However, this statement ignores the strong competition that both root crops and cereals face from vegetables and other horticultural crops, the subject of the succeeding chapter.

FURTHER READING

D. K. Britton, *Cereals in the United Kingdom* (1969)

B. E. Cracknell, *Past and Future Cereal Production in the United Kingdom* (1970)

F. G. Hay, *Report on the Marketing of Scotch Seed Potatoes* (n.d., ?1969) Univ. of Glasgow

J. D. Sykes and J. B. Hardaker, *The Potato Crop* (1962)

CHAPTER 13

Horticulture

Flowers, fruit, nursery stock and vegetables are generally considered separately from other crops, under the heading of horticulture, and are estimated to account for some 10 per cent. of gross agricultural output in the United Kingdom. As their name indicates, horticultural crops were originally grown in gardens, generally close to towns, and this usage survives in the term 'market garden'; but the distinction is less clear-cut today when vegetables and, to a lesser extent, flowers and soft fruit may be grown as rotational crops on arable farms. From the late nineteenth century, such crops were gradually adopted by arable farmers as additional cash crops, a practice greatly extended since 1919. R. R. W. Folley defined horticulture as high-investment crop production 'intended for human use without prior conversion into a different type of edible product', a definition which excludes hops and sugar beet, but makes no reference to location or to the kind of producing unit; he also omitted potatoes as being conventionally regarded as farm crops and Miss Ruth Gasson's suggestion that 'horticulture' be replaced by the term 'intensive crops' would include potatoes, but not hops and sugar beet. There are arguments for both approaches but, while it may be illogical not to regard potatoes as a vegetable, they are grown more widely and on a far larger scale than other 'intensive crops', and Folley's definition is both consistent with current practice and conforms with official usage, which treats horticultural crops separately for statistical analysis and government assistance. The term horticulture will therefore be retained.

Perhaps the most distinctive feature of horticulture is its variety, and the report of a joint survey of the industry by government departments and the farmers' unions has expressed doubts whether many generalisations about horticulture as a whole are valid. It not only includes numerous crops, but these vary in their physical requirements, the scale and trend of production, the enterprises with which they are associated, their susceptibility to foreign competition, their contribution to horticultural output, and in many other respects. Nor can they conveniently be divided into discrete categories, for the same crop may be grown at different scales and by different methods; thus, asparagus is grown both as a small-holder's crop in Worcestershire and on a large scale on Breckland arable farms, while crops for canning and quick-freezing are increasingly produced in different ways from those intended for other markets. This variety makes it difficult both to estimate the comparative return from different crops and to study the geography of horticultural production as a whole. Here it will be convenient to consider horticultural crops under three heads: outdoor vegetables and flowers; fruit; and glasshouse crops. These account respectively for 49 per cent. (12 per cent. flowers), 26 per cent. and 25 per cent. of horticultural output, and 63 per cent. (5 per cent. flowers), 37 per cent. and under 1 per cent. of the acreage under horticultural crops.

CHARACTERISTICS OF HORTICULTURE

Some general characteristics of horticulture should first be noted. Horticultural crops are associated with a high standard of living, and the emergence of modern horticulture dates from the nineteenth century, when urbanisation, population growth and rising standards created a demand for fruit and vegetables. Demand has continued to grow, but the products demanded have changed; for while the income elasticity of demand for horticultural produce (i.e., the percentage increase in demand for each 1 per cent. increase in income) is generally high, it is higher for the choice products, such as salad crops, which are increasingly demanded as incomes rise. Moreover, as the market becomes more sophisticated, consumers become more discriminating in respect of quality, appearance and freshness, and the prices individual growers receive for the same product consequently vary widely, despite similar costs of production. Discrimination in the demand for horticultural crops is generally much greater than in agriculture and the rewards for success and penalties for failure correspondingly larger. This characteristic helps to make horticulture a highly competitive industry and encourages the search for the most favourable locations.

Like dairy farmers, horticulturalists have faced increasing competition from overseas supplies; for, while the natural protection from imports, still enjoyed by rather more than half the home-grown produce because of its perishability or bulk, was one consideration which encouraged expansion of horticulture since the late nineteenth century, the improvements in transport which, by reducing costs and speeding deliveries, extended the area within Great Britain in which horticultural crops could be grown profitably, also facilitated competition from areas where natural conditions were more favourable for fruit and vegetables, such as the Mediterranean coastlands of Europe, or horticulture was organised more efficiently, as in the Netherlands. The home-grown crop often has advantages in freshness and flavour, which is frequently better where growth is slow, to offset the lower costs of production of imported produce, but imports of temperate products directly competitive with British produce account for 21 per cent. of all horticultural imports. Nearly half the imports are tropical and sub-tropical crops which cannot be grown in Great Britain. The volume of imports would undoubtedly be larger if British horticulture were not sheltered to some extent by tariffs and quotas, often varied seasonally to give added protection to the home crop; for example, only 18 per cent. of the quota of apples from non-Commonwealth countries may be imported between July and December, when most of the home crop is being marketed.

A further 31 per cent. of imports consists of temperate products which arrive outside the main season, for rising standards have led to a growing demand for a regular supply of a wide range of horticultural products, a far cry from the situation two generations ago, when the British public ate fruit 'during the season and in the quantity ordained by our local climatic conditions'. While the climate of the British lowlands, with its long growing season and infrequency of severe frosts, is nearly ideal for certain kinds of vegetables, supplies of which are available throughout the year, most products grown in the open are harvested only in summer and autumn; some 73 per cent. of British horticultural produce is grown during the main season, compared with 14 per cent. of early produce and 13 per cent. out of season. An increasing quantity of this produce is being preserved by canning, quick-freezing or dehydration, or by keeping in cold storage, to be marketed over a longer period, but the proportion is still smaller than that marketed fresh; much out-of-season produce is therefore imported.

To varying degrees, horticultural crops exhibit three common characteristics: they are produced by intensive methods; their yield, and hence profitability, is highly variable; and they are more highly localised than most agricultural enterprises. Many horticultural crops are grown on small acreages with the application of much hand labour. The total area under horticulture, according to the June census, is only about 620,000 acres and, while this somewhat understates the acreage of annual crops by omitting both early vegetables which have already been harvested and late crops which have not yet been sown, the area under horticultural crops at any one time represents only about 5 per cent. of that under tillage. Yet the share of the labour force engaged in horticulture is 12 per cent., for labour requirements for most horticultural crops grown in the open range between 17 and 100 man-days per acre, compared with between $2\frac{1}{2}$ and 20 for most farm crops, and up to 1,500 man-days for glass-house crops. Moreover, the decline in the horticultural labour force has been less steep in recent years. Horticulture also makes extensive use of seasonal and part-time workers, especially for tasks such as picking; women, who generally possess greater manual dexterity, are particularly important in such work and the proportion of female labour is high. Except in cultivation, mechanisation of horticultural crops has made only limited progress, especially with the choicer products, which must be handled carefully, and it generally results in lower yields, although output per man may be much higher; picking may account for half the cost of producing crops such as raspberries and runner beans, and a machine developed for harvesting French beans works at a rate equivalent to 150 pickers. Other forms of labour-saving have made more progress, and the development of pre-emergence herbicides has reduced by four-fifths the labour required for cleaning, which rarely accounts for less than 10 per cent. of the labour bill. Nevertheless, the labour requirements of horticulture remain high, so that labour is by far the largest input on horticultural holdings, accounting for 40 per cent. of inputs (other than seeds, livestock and feeding stuffs) on such holdings in the Farm Management Survey. On the other hand, capital investment per man is also generally higher than in farming, notably in glasshouse crops, and the average output per acre is over eight times higher than from all farm crops. Since less is spent on outside purchases than in farming, the ratio of net outputs is even more favourable. It is, therefore, undesirable to examine only acreage changes, especially over recent decades, when yields have been rising and a declining acreage may be accompanied by an increase in production; unfortunately, there are no county estimates of production available.

Although output per acre of horticultural crops is high, their yields vary more than those of many farms crop and, except for crops grown in sheds and under glass, both yield and quality are much affected by unfavourable weather, especially at critical phases such as pollination of fruit blossom; as an extreme example, the 1935 apple crop was only 18 per cent. of that from virtually the same acreage in the previous year. Since short-term demand is fairly constant and costs of production remain unchanged, the value of the crop may vary greatly from year to year; thus, the co-efficient of variation in value between 1956 and 1965 ranged from 9·9 per cent. for wheat to 23·8 per cent. for carrots. Stability of output is not helped by the extent to which farmers with only a minor interest in horticulture grow crops as a speculative venture. This variability in output explains why governments support growers, not by guaranteed prices as with farm crops, but through tariffs and quotas on imported produce.

While small acreages of horticultural crops are widely grown to satisfy local demands, horticultural production is more highly localised than that of cash crops in general (Figure 20), and is overwhelmingly concentrated in southern England. Scottish horticulture provides only 4 per cent. of agricultural output in Scotland and only 4 per cent. of the horticultural total for Great Britain, whereas England south of the Wash-Mersey line provides over three-quarters; indeed, nearly two-thirds of production is from the Eastern and South-East Regions which have a broadly similar share of each main class of horticultural produce, although flowers, glasshouse produce and vegetables are rather more widely dispersed and fruit less so (Table 31). Kent, the Fenland, mid-Bedfordshire, East Anglia and the Evesham area are the principal localities, with often a surrounding zone of lesser intensity where a small range of crops is grown. The location of some individual crops is even more marked, particularly where the total acreage grown is small; for example, 75 per cent. of the celery grown in 1961 was

TABLE 31
DISTRIBUTION OF HORTICULTURAL OUTPUT IN GREAT BRITAIN, 1965-6

Region	Total Output %	Vegetables %	Fruit %	Flowers %	Glasshouse Produce %
Eastern	38	41	28	44	43
South-east	22	12	38	11	22
W. Midlands	10	10	16	3	3
South-west	5	5	15	14	4
Other	25	32	13	28	28
ALL	100	100	100	100	100

Source: W. C. Hinton, *Outlook for Horticulture*

produced within 20 miles of Littleport (Isle of Ely) and over 50 per cent. of the Brussels sprouts within 15 miles of Biggleswade (Bedfordshire). Such localisation often arises by chance or from the initiative of a single grower; but, once established, it tends to persist, given favourable conditions, by virtue of reputation, market connections and the development of a specialised labour force and ancillary industries. Such advantages may enable production to continue even where the circumstances originally stimulating its development no longer apply; thus, mid-Bedfordshire remains a major centre, although the railway connection which was largely responsible for its growth in the nineteenth century, with two-way traffic of vegetables and London manure, is now unimportant. Heavy capital commitment accentuates this inertia, and the Lea Valley retains the largest acreage of glasshouses in the country 'despite all economic logic.' Nevertheless, the location of horticultural production and the relative importance of different areas do change. In the nineteenth century, encouraged by the railways and later by the depression in arable farming, production of horticultural crops spread from its long-established locations near towns to more distant areas where soil or climate was favourable; since 1920, there has been a further shift towards eastern and south-east England, initially because of a faster rate of growth in these areas and later, especially with increasing competition since the Second World War, because of a contraction of production in western districts and an increasing concentration in those eastern areas judged to enjoy the greatest comparative advantage. This move has been facilitated by the switch from rail transport, for which some expanding areas were not well placed, to road and by the widening range of horticultural crops which farmers can grow.

The difficulty of providing satisfactory explanations for the distribution of horticulture is often acute. It may be possible to indicate the necessary conditions for the location of horticultural production, although these are generally inferred, rather than demonstrated; it is often hard to define the sufficient conditions which explain why, of all possible locations, these were selected.

OUTDOOR VEGETABLES & FLOWERS

Outdoor vegetables occupy some 59 per cent. of the acreage under horticultural crops, and 37 per cent. of output by value, flowers 5 per cent. and 12 per cent. These figures indicate the less intensive nature of much vegetable production, an impression confirmed by the estimate that field vegetables employ only 22 per cent. of the labour devoted to horticultural crops. Figure 19 shows the distribution of the acreage under vegetables in 1965. It is overwhelmingly concentrated in eastern England, especially the Fenland, with small outliers in south-west Lancashire, the Evesham district and west Cornwall, although the use of a symbol representing 1,000 acres masks the widespread scatter of small acreages; in Scotland, which has only 3 per cent of the acreage, production is located mainly in the Lothians and around Dundee. The distribution of flowers and nursery stock more nearly resembles that of the urban population, although the Fenland and south-west England are also prominent.

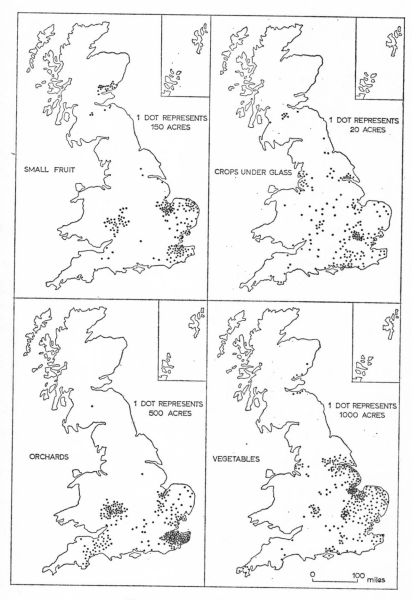

Figure 19. Horticultural crops
Source: agricultural censuses 1965

As with other crops, the proportion of holdings growing vegetables varies both nationally and regionally with the importance of the crop. Vegetables are recorded on only 11 per cent. of holdings in England and Wales, but the proportion is 31 per cent. in Eastern England and 2 per cent. in Wales; in Scotland, 3 per cent. of the holdings have vegetables but in the Highlands only 1 per cent. Unlike most agricultural cash crops, vegetables are most frequently grown on both large holdings (500 acres and over) and small (under 50 acres), although the proportion of the total acreage grown in both size groups varies greatly; the chief exceptions are crops, such as peas for harvesting dry, which are well suited to mechanised methods. Table 32 shows that the proportion of the vegetable acreage on holdings of 300 acres and over ranges from 74 per cent. in the East Midlands (which includes Lincolnshire, a major county for large-scale vegetable growing on arable farms) to 3 per cent. in Northern England, and that on holdings of under 50 acres from 7 per cent. to 49 per cent. respectively in the same regions. There are similar contrasts in the acreages grown, though large holdings do not necessarily have large acreages under vegetables. Thus, 51 per cent. of the vegetable acreage in the East Midland Region is on holdings with 100 or more acres of vegetables, compared with only 6 per cent. in Northern England, the relationship being reversed on holdings with less than 10 acres of vegetables, which have 5 per cent. and 44 per cent. of the vegetable acreage respectively. However, as Table 33 shows, there is a marked trend for the proportion of vegetables grown in large acreages to increase in response to the demand for crops for processing.

The area under vegetables, although two-fifths larger than before the Second World War, declined steadily from the late 1940s, primarily because of a great fall in the acreage of peas harvested dry, but also because of declining acreages (although not production) of most vegetable crops; but this decline has been arrested. These trends have not been uniform: the popularity of some crops, such as savoys, has declined, while that of others, such as lettuce, has risen; peas and dwarf beans for quick freezing are exceptional in showing a marked increase in acreage. With more than twenty major vegetables and several different kinds of bulbs and cut flowers, each with specific requirements, it is often misleading to discuss vegetables in general terms and impossible to consider each within the confines of this book; fortunately, a full account of the distribution of vegetables in England and Wales was published by

TABLE 32
VEGETABLES IN THE REGIONS OF ENGLAND AND IN WALES, 1968

	Eastern	South-east	East Midlands	West Midlands	South-west	Yorks and Lancs	Northern	Wales
	Percentage of holdings with vegetables							
	31	13	11	8	7	12	3	2
	Percentage of vegetable acreage in England and Wales							
	50	10	17	6	4	11	1	1
Holdings (crops and grass)	Percentage of vegetable acreage on large and small holdings							
300–	57	42	74	21	17	40	3	12
–50	13	20	7	38	34	18	49	39
ALL	100	100	100	100	100	100	100	100
Vegetable acreage groups	Percentage of vegetable acreage by vegetable acreage groups							
–10	10	15	5	26	34	13	44	42
10–49	31	35	22	42	50	43	40	34
50–99	23	21	22	13	10	19	10	6
100–	36	29	51	18	6	25	6	18
ALL	100	100	100	100	100	100	100	100

Source: Ministry of Agriculture

TABLE 33
VEGETABLE ACREAGE GROUPS IN ENGLAND AND WALES, 1960 & 1968

Acreage Size Groups	1960		1968	
	Growers %	Acreage %	Growers %	Acreage %
–20	88·6	36·5	81·7	23·2
20–49	7·8	24·9	10·8	21·9
50–99	2·4	17·3	4·6	20·6
100–	1·2	21·3	2·9	34·3
ALL	100·0	100·0	100·0	100·0

Source: Ministry of Agriculture

the Ministry of Agriculture in 1967. Some 45 per cent. of the vegetable acreage in England and Wales and 35 per cent. of that in Scotland carried pulses (mainly peas for harvesting dry and for canning and quick freezing), 40 per cent. and 33 per cent.

respectively brassicas, and 13 and 16 per cent. root crops. The main distinction is between coarse vegetables, such as cabbage, grown on a field scale, and the choice vegetables, planted generally in smaller acreages and occupying a much smaller proportion of the land devoted to vegetables.

Production of coarse vegetables is generally more widely dispersed, although there are some remarkable concentrations, including that of Brussels sprouts (already noted) carrots (45 per cent. in Norfolk), green peas for processing (35 per cent. in Lindsey) and summer cauliflower (32 per cent. in Holland). Small acreages of choice vegetables are also widely scattered, but production is often more localised, both within regions and within the country as a whole. Flowers and bulbs generally exhibit many of the features of the production of choice vegetables. It should, however, be noted that, even for the most widely grown vegetables, the proportion of holdings with any particular crop is small: only 3·6 per cent. of the holdings in England and Wales return cabbages and only 0·4 per cent. early carrots.

FACTORS INFLUENCING THEIR DISTRIBUTION

It is tempting to look at the distribution of vegetable- and flower-growing in terms of natural advantage and indeed most of these crops are grown in areas which are well suited to growing crops in general. Large quantities of vegetables are grown in all the principal areas of cash cropping, except on the chalk downs of southern England, where soils are too thin; the acreages grown in Scotland and northern England are also comparatively low. Vegetables are generally grown on medium to light soils and are prominent on nearly all the first class (A) land, although this is in fact a circular argument, because one reason for an A classification was the large proportion of farmland under horticultural crops. Only a small acreage of vegetables is grown on heavy land or stony soils, such as those derived from clay-with-flints, for ease of working is a major consideration. On the other hand, some areas of light soils, which once depended on abundant manure from the towns, have problems of maintaining soil organic matter since these supplies ceased. For some crops, especially roots and celery, soils are a major consideration. Carrots, for example, tend to split when grown on heavy land and are difficult to clean, a matter of some importance now that perhaps a sixth are grown for processing and many more are pre-packed. Large acreages of carrots are grown on the light soils of north-west Norfolk, though this area was not the first choice; for production spread there from the peat Fens (where it was already established), when large growers were unable to obtain sufficient land. For other crops, soils matter less; asparagus, for example, is grown on both light soils in the Breckland and heavy soils in Worcestershire, and could be much more widely grown.

Climatic conditions suitable for cash crops are generally satisfactory for vegetables, but high output and variable yields from vegetables can more often justify investment in spray irrigation; about 40,000 acres of vegetables were irrigated in England and Wales in 1965, two-fifths of it in the Eastern Region (although Worcestershire had the highest proportion and the largest acreage of any county). Climate determines to some extent the range of crops that can be grown in the open and is of special importance for the production of early crops and of those, such as cauliflower, which are sensitive to frost. Coastal sites in Cornwall, with sea-temperatures in winter some 4–5° F. higher than on the North Sea coast, have considerable advantages, with a higher intensity of radiation, a lower risk of frost and the highest accumulated temperatures in January and February ($+250°$ F.). Early vegetables and flowers are prominent among the horticultural crops of such areas, although much early produce is grown in the main vegetable areas, often on specialist holdings. Nevertheless, many other localities could grow early horticultural crops, and for some crops, such as Brussels sprouts, frost is even beneficial.

Undoubtedly, these physical factors provide part of the necessary conditions for the production of vegetables and outdoor flowers, but it will help an understanding of the distribution of such crops if the evolution of the present pattern of vegetable growing is briefly outlined. As already noted, vegetable growing was first located around towns, with small concentrations in more distant areas, such as Sandwich and Sandy. The railways greatly stimulated production in such outlying areas, and also played an important part in the development of areas of early production such as west Cornwall. These areas were often, though not invariably, favoured by soil or climate, and speedy access to urban markets, the only outlet, was important for most crops. Depression in arable farming from the late nineteenth century encouraged the extension of market gardening into agriculture, when arable farmers began to grow vegetable crops, especially those that they were already producing as stock feed, such as cabbages and turnips. The range of crops was gradually extended and it is claimed that the expansion of sugar-beet growing in the 1920s gave farmers the necessary experience to grow tap-rooted vegetables. At the same time, production was also expanding in the established districts of specialised vegetable growing and farmers in adjacent areas were often among the first to adopt vegetables as a farm crop, as with the growing of Brussels sprouts in north Hertfordshire. This extension to new areas was greatly helped by the advent of the motor lorry and, to a lesser extent, by the establishment of canneries in eastern England in the 1920s and 1930s. Thus, between 1921 and 1936, the acreage of vegetables grown in Norfolk increased by 327 per cent. and in Suffolk by 269 per cent., compared with an average national increase of 80 per cent.

Since the 1930s no other major centre for growing vegetables has emerged, but production has continued to rise faster in eastern counties in the post-war period, particularly those surrounding the Wash, and these changes must be seen in the context of a declining national acreage under vegetables and rising national yields. Thus, while the acreage under vegetables in Holland, Lindsey and Norfolk increased by 29 per cent. between 1956 and 1967, their share of the vegetable acreage in England and Wales (which fell by 11 per cent.) rose from 27 per cent. to 39 per cent., partly at the expense of long-established areas such as the Evesham district. Concentration was aided by the widespread adoption in the early 1950s of large diesel lorries for transporting vegetables and by the local markets created by the building of quick-freezing plant (especially in east-coast fishing ports) for major crops such as peas and carrots. A similar trend towards concentration of production occurred in nearly all the major vegetable crops, although the leading crop is not always the same; thus, Lindsey's share of green peas for processing rose from 9 per cent. to 35 per cent. at the expense, relatively, of Norfolk (27 per cent. to 19 per cent.), and Norfolk's share of carrots rose from 29 per cent. to 45 per cent. at the expense of the West Riding and Lindsey (23 per cent. to 8 per cent.).

This expansion and subsequent concentration of vegetable growing in arable areas has thus led to a dichotomy between the general farmer, growing vegetables as a side-line (although often an important one) and the market gardener, for whom horticultural produce provides the bulk of farm income. According to W. L. Hinton, farmers produced 68 per cent. by value of the field vegetables in 1965–66 and 50 per cent. of the field flowers, the balance being grown by market gardeners. In practice, the distinction is not easy to draw. Neither size of holding nor acreage of vegetables is a very satisfactory criterion, for a small acreage may be grown as a speculative crop on a small holding which does not specialise in vegetable growing, while there are market gardens of several hundred acres. In general, market gardeners have smaller holdings than farmers, although these may range from a few acres to over 300, and much of their land is under horticultural crops (though only some will be vegetables); they tend to grow a larger number of vegetables and concentrate on the choice vegetables, which require more attention and hand labour. In general, the larger the holding, the lower the percentage of intensive crops, and where the commoner vegetables are grown, they are likely to be earlies. Farmer-growers have vegetables or flowers as only one of several cash crops, and vegetables occupy a small proportion of the holdings; larger acreages of only one or two vegetables are grown, generally coarse vegetables that are easily mechanised. However, the distinction between the two kinds of producers is complicated by the trends for all vegetable to be grown on a larger scale, for the range of vegetables grown by farmers to increase and for that grown by market gardeners to decrease.

Growing vegetables on farms has been encouraged by several advantages. The crops are often those such as cabbage, carrots and peas, which have low labour requirements and can be mechanised more easily; and because the farmer specialises in only one or two crops, he can acquire expertise. Growing field vegetables may also offer economies of scale, and this is among the most efficient sectors of the horticultural industry. Furthermore, such vegetables often fit conveniently into the farm economy in various ways. They serve as break crops which are also cash crops and take the place of traditional cleaning crops. They may make more efficient use of labour and possibly of machinery by spreading the demand for men and machines more evenly throughout the year; thus the harvesting of peas precedes that of grain and Brussels sprouts are picked after cash roots have been lifted. Some crops, such as peas for harvesting dry, also require little additional equipment. Large producers are preferred by processors, who sometimes prescribe minimum acreages and let contracts for substantial acreages to a comparatively small number of farmers. Lastly, because they depend less on vegetables, farmer-growers are less vulnerable to price changes, for they can feed a crop to livestock (turnips and swedes, for example) or plough it in as green manure, thereby avoiding the expense of harvesting, a major cost of production. For the market gardener there are few economies of scale in the crops he grows, for most have heavy labour requirements, whether for cultivation, as with beetroot, or in picking, as with runner beans, or in preparation, as with leeks.

However, the kind of external economies which arise from the areal concentration of market gardening and vegetable-growing generally should not be overlooked. An example is provided by merchant-growers in the Fenland, substantial growers who also organise production by others, buy the crop at an agreed contract price and provide equipment and skilled gangs for harvesting, thus giving the smaller growers many of the advantages of large-scale production. Co-operative marketing and market links play a similar role; the localisation of the production of individual crops is maintained and strengthened where such channels exist, as with the specialised marketing of swedes from south Devon.

Markets, whether in the towns or in processing plant, also play a part, though their location is generally less important than formerly (Chapter 6). Differences in the demand for vegetables throughout Great Britain also appear to exercise some influence and the low per capita consumption of green vegetables in Scotland may help to explain the comparatively small acreage of vegetables grown there.

The existence of a large pool of experienced seasonal workers with a tradition of horticultural work is also an advantage; the growing of carrots in the Fenland might not have developed without such a pool, and its absence perhaps explains the decline of celery-growing in the Trent lowlands. These manmade advantages reinforce natural advantages and encourage the further concentration of vegetables as their acreage declines, although, as already noted, this trend is not peculiar to horticulture, but only well exemplified by it.

ORCHARDS & SOFT FRUIT

Fruit of all kinds provides some 26 per cent. of horticultural output and occupies about 37 per cent. of the acreage under horticultural crops. It is generally more localised than vegetables, although orchard fruit and soft fruit, which share the fruit acreage in the proportion of four to one, differ in this respect (Figure 19). Since their requirements and characteristics are different, they will be treated separately.

Orchards are largely restricted to southern England; there are only about 700 acres in Scotland, 1,200 acres in the two northern regions of England, and 1,500 acres in Wales (Figure 19). Even this distribution gives a misleading impression of the location of commercial fruit growing, for it includes numerous small farm orchards which are often not very productive. Since 1964, however, the acreage of fruit grown commercially has been separately recorded. These commercial orchards, accounting for 79 per cent. of the acreage under orchards in England and Wales (though only 43 per cent. of the holdings with orchards) are mainly located in eastern England, notably in Kent (42 per cent. of the total acreage) and in East Anglia (21 per cent.), where production is more dispersed; most of the remainder are in Gloucestershire, Herefordshire and Worcestershire (18 per cent.) and in Somerset and Devon (6 per cent.). Non-commercial orchards, on the other hand, are primarily to be found in these western counties, which have over 60 per cent. of the acreage.

The distribution of orchards is more localised than the map

suggests, especially in Kent, where most orchards are situated on the High Weald, the escarpment of the Lower Greensand around Maidstone and the dip-slope of the North Downs; furthermore, only 4 per cent. of the holdings in England and Wales have commercial orchards. Unlike vegetables, the proportion of holdings with orchards changes little from one size group to another: 20 per cent. of the acreage of commercial orchards is on holdings with under 50 acres of crops and grass and 28 per cent. on holdings with 300 acres and over. There is, however, a marked contrast in the acreage distribution of commercial and non-commercial orchards (Table 34): while

TABLE 34
ORCHARDS IN ENGLAND AND WALES, 1966

Orchard acreage Size groups Acres	Commercial Holdings %	Orchards Acreage %	Non-Commercial Holdings %	Orchards Acreage %
−10	72·3	15·8	95·6	69·7
10–19¾	11·0	10·5	3·3	15·6
20–49¾	9·8	21·5	1·0	10·5
50–	6·9	52·5	0·1	4·2
ALL	100·0	100·0	100·0	100·0

Source: Ministry of Agriculture

96 per cent. of the latter are under 10 acres and these account for 70 per cent. of the acreage under non-commercial orchards, over half the commercial acreage is on holdings with 50 or more acres of orchards. This contrast is presumably increasing as the proportion of the commercial acreage on large holdings rises.

FACTORS INFLUENCING THE DISTRIBUTION OF ORCHARD FRUIT

As this concentration of commercial orchards in eastern and south-east England suggests, climatic considerations are of major importance, although the advantages are relative. No part of the country is ideal for orchard fruit; average yields of apples in Italy are twice as large as those in Great Britain, though differences in quality offset some of the advantages of higher yields. Within Great Britain, high rainfall increases the risk of disease, and the quality of fruit grown in wetter areas is generally inferior; wind is also a major hazard. Warm, sunny weather is desirable when the fruit is ripening and these conditions occur most frequently in eastern and south-east England; irrigation is now used to increase yields in these drier areas and about 10 per cent. of the commercial acreage in East Anglia and about 4 per cent. of that in Kent were irrigated in 1965. Early onset of growth is not as important as in vegetable growing, but late spring frosts are a major cause of variable yields; coastal sites are generally too exposed, but the risk of frost elsewhere can be minimised by careful siting, although many orchards, even in Kent, are not well-sited, a fact of some significance since Bramley Seedling, the major variety of cooking apple, is particularly sensitive to frost. The principal soil requirements are that it should be fairly deep and well-drained, so that a good root system can develop, although the different fruits vary considerably in their tolerance of poor drainage, with cherries the most sensitive and plums and cooking apples the least. One other natural advantage of Kent is the large acreage of soils that satisfy these conditions, especially those developed on the fine-grained sands of the High Weald, the Ragstone of the Lower Greensand and the plateau loams and brick earths of north Kent. The physical suitability of sites for fruit growing is becoming more important as an increasing proportion of the orchard acreage is devoted to dessert apples and pears, which must be of good quality and appearance to compete with imported produce. Poorly-sited orchards are unlikely to satisfy these requirements.

Such physical considerations provide only a partial explanation of the localisation of top-fruit growing. The way the industry has evolved and is evolving is equally important, both because top fruit represents a long-term investment and because reputation for fruit growing has played an important role. Orchards may involve an investment of £500 per acre before any return is forthcoming and the economic life of orchards may be as much as 50 years. Unlike vegetables, fruit trees are rarely a speculative venture and, once in bearing, they are likely to remain even when the circumstances that brought them into being have changed. Kent, through its proximity to London and the Continent (from which much early knowledge of fruit-growing came), has long been the premier county for top fruit, and its reputation, since the sixteenth century at least, as the 'Garden of England' has undoubtedly contributed to its long pre-eminence. In the mid-nineteenth century, Kent and Middlesex in eastern England and Devon, Gloucester, Hereford and Worcester in the west were the principal areas for orchard fruit, the latter group mainly because of their large acreage of cider apples and perry pears. The rapid expansion of London destroyed many orchards, which were often replaced by new plantings in Kent, for there was little suitable land in other nearby counties. At the same time, the rapidly increasing urban population and the prolonged depression in arable farming stimulated both new plantings in established areas, such as Kent and Worcestershire, and also the emergence of new areas, as around Wisbech and Cambridge and in south Essex. The character of the orchards was also changing, and the dominance of cider orchards (estimated at over 100,000 acres in 1913) declined as new orchards of apples, cherries and plums were planted. Until the First World War, the acreage under orchards increased fairly steadily, although this may partly reflect better enumeration; thereafter, with total acreage changing little, eastern districts became steadily more important, not only because most of new planting was concentrated there, but also because large acreages of older orchards in western counties were being grubbed up and not replanted. Thus, while the area under orchards in Devon, Hereford, and Somerset fell by more than 10 per cent., the acreages in Essex and Suffolk rose by over 100 per cent. and in Norfolk and Sussex by more than 60 per cent. (compared with 8 per cent. in England and Wales as a whole). Kent's acreage also rose by 81 per cent. to give that county 26 per cent. of the 260,000 acres in England and Wales. This trend largely reflected the increasing commercialisation of fruit production and the search for alternatives to traditional arable farming.

After the Second World War (which had little effect on the distribution of top fruit), the trend continued; but, from the early 1950s, it represented a relative, not an absolute, change, for the acreage under orchards began to fall steadily, though less steeply in eastern counties than in western. This fall does not imply a decline in the importance of top fruit, for yields have been rising, the total volume of fruit produced is greater and a substantial acreage of new orchards has been planted; it is also due to the removal of old, uneconomic orchards, a process helped by government grants and reinforced by the Horticultural Improvement Scheme, begun in 1960. For example, some 30,000 acres were planted with top fruit between 1951 and 1957, but 60,000 acres were grubbed up. Because planting and removal are not uniformly distributed, the share of eastern and south-eastern counties has increased; thus, Kent's share rose from 26 per cent. in 1951 to 36 per cent. in 1966, although the orchard acreage in the county declined by 16 per cent. (compared with 33 per cent. in England and Wales). Since most of the orchards destroyed were of low productivity, while the new orchards consist of improved varieties and are generally well-managed, the county's share of top fruit production (allowing for the lag of up to ten years before new orchards are in full bearing) has probably increased even more markedly. Many of these new orchards are under dessert apples and pears, a crop some three times as valuable as cooking apples and more than seven times as valuable as cider apples, and for which east and south-east England have a greater comparative physical advantage than for other orchard fruit (except cherries). Thus, only 23 per cent. of the orchards of dessert apples and 24 per cent. of those of dessert pears were more than 25 years old at the 1966 fruit census, compared with 73 per cent. of cooking apples, 58 per cent. of cherries and 48 per cent. of plums, proportions which would be considerably higher if large acreages had not been grubbed up. As a result, the decrease in the average acreage of dessert apples was only 2 per cent. between 1952–6 and 1964–6, compared with 30 per cent. for cherries, 34 per cent. for cooking apples, cider apples and perry pears, and 43 per cent. for plums. Rising yields have offset the decline in the fruit acreage in the same period, and the output of dessert apples increased by 65 per cent. and that of cooking apples by 14 per cent.; the yields of cherries and plums, on the other hand, declined and output has fallen even more sharply. This is a further example of the danger of considering only changes in acreage.

The progressive concentration of fruit growing in south-east and, to a lesser extent, eastern England, has been assisted by many of the factors that promoted concentration in vegetable growing, although little top fruit, other than plums and cider apples, is processed, and the location of processing plant is not a major consideration. Concentration of fruit-growing also promotes the development of specialised marketing facilities; for example, a packing station requires at least 500 acres of trees in full bearing within a radius of 10 miles if it is to be operated efficiently. Co-operative marketing is increasingly common among fruit growers and one group claims to market a quarter of all home-grown dessert apples. Similarly, an increasing proportion of top fruit is being kept in cold stores to extend the marketing season and nearly a third of the apple crop is probably marketed in this way. There are over 1,000 such stores and, where they are co-operative or shared ventures, they are more likely to be erected in the principal fruit-growing areas. Research stations and specialist advisory services also tend to be located there. Like vegetable growing, fruit production is helped by a large, experienced force of both full-time and part-time workers. Labour still accounts for about half the cost of production (if the manual labour of growers is included) and it has been suggested that skills in fruit growing take longer to acquire. Harvesting is highly concentrated in October, when labour requirements may be four times those of other months, but packing continues until February and pruning also employs skilled seasonal labour, which accounts for some two fifths of labour costs on fruit farms in south-east England.

There appear to be few economies of scale in growing top fruit, which does not lend itself to mechanisation, except for carting, spraying and the like, and there is evidence that orchard crops are not well suited either to the small grower, who may have problems of pest control and marketing, or to the very large holding, where it may be more difficult to ensure high quality. As a result, the dichotomy noted in vegetable growing is less marked; 60 per cent. of the output of all fruit is grown by specialists, 24 per cent. by farmers and 16 per cent. by market gardeners, and these figures must broadly indicate the proportions of orchard fruit, since this accounts for 68 per cent. of all fruit by value.

SOFT FRUIT

The production of soft fruit is smaller in scale and more dispersed than that of top fruit. It is most prominent in the main areas of commercial orchards, but soft fruits are also widely grown in East Anglia, especially on the loams of east Norfolk and on the Fen silts, and production extends much further north, with nearly a quarter of the acreage in Scotland, mainly in Perth and Angus. Climatic requirements are less exacting than for orchard fruit, except for early strawberries, where prices can halve between the end of May and late June; a site with a favourable aspect along the south coast of south-west England was highly advantageous for strawberry growing before the advent of cloches, and crops from this area are still the first of the unprotected crops to be marketed.

The growing of soft fruit has seen a similar extension from market gardening into arable farming, chiefly blackcurrants, which have a life of several years and were widely adopted as a rotational crop on arable land in East Anglia in the 1920s and 1930s; gooseberries, raspberries and strawberries, on the other hand, are more likely to be grown on small holdings. Labour costs are high, especially for picking, although these crops offer greater prospects of mechanisation than orchard trees, and some blackcurrants are already being harvested mechanically, a practice which involves biennial harvesting and so a lower yield, but which greatly reduces labour costs. But perhaps the market is the most important influence on the location of soft-fruit growing, whether for consumption as fresh fruit or for processing. Raspberries and strawberries are perishable and speedy access to urban markets is highly advantageous; transport represents between a fifth and an eighth of marketing costs, a higher proportion than for most products. About half the strawberries and most other fruits are processed, and much of the acreage under soft fruit is located near canneries, quick-freezing plant and jam factories, although some of these were attracted to such areas because soft fruit was already being grown there. Nearly all blackcurrants are grown on arable farms on long-term contracts for processors, whose location helps to explain much of the soft-fruit growing in the West Midlands (although there is also a long tradition in Worcestershire). The raspberry crop in Angus and Perth, one of the most highly localised and specialised of all the horticultural areas, is also largely processed. Growers have an average of 8 acres each, compared with less than 1 acre in England and Wales, where much of the crop is consumed fresh and production is more widely dispersed on small holdings, many of them serving local markets.

As with other horticultural crops, the acreage under soft fruit is falling, although the decline began much earlier, in the 1930s. It fell by a further 6 per cent. between 1952–6 (3 year average) and 1964–7, when only the acreage under blackcurrants increased, although, with higher yields, the production of blackcurrants rose 35 per cent. and of strawberries 20 per cent. However, with the exception of raspberries, where the proportion grown in Angus and Perth rose from 64 per cent. in 1956 to 76 per cent. of a smaller acreage in 1967, no general pattern of localisation emerges.

Unlike orchard fruit, soft fruit is rarely a major enterprise on holdings and is more generally associated with other fruit than with vegetables or general farm crops.

CROPS UNDER GLASS

Crops grown under glass provide a quarter of horticultural output and represent the most intensive form of cultivation, having an average gross output of over £5,000 per acre, compared with £250 for all horticultural crops, and requiring 1,500 standard man-days per acre. They account for less than 1 per cent. of the acreage under horticultural crops, yet employ some 27 per cent. of the labour used in horticultural production. Glasshouse cultivation is also a most interesting branch of horticulture, for it presents several paradoxes. Glass requires the heaviest investment per acre of all kinds of crop production, of the order of £25,000 per acre, yet it is more flexible than top-fruit growing, another high investment crop, and there have been considerable changes in the crops grown under glass. Secondly, although crops in glasshouses are grown in artificial soils and artificial climates, 'climatic factors probably have a greater effect than in any other section of horticulture or agriculture' (L. G. Bennett). Thirdly, it exemplifies the conflict between optimum physical location and the pull of the market.

There are some 4,150 acres under glass, 90 per cent. of it in England. Figure 19, which records the main concentrations, shows that glasshouses are widely distributed, with the largest acreage in the Lea Valley (about 13 per cent.), and with other important areas around Blackpool and Hull, along the coast of Sussex and in the Clyde Valley, which has half Scotland's 300 acres of glass; many of the remainder are associated with urban areas. Like many other horticultural enterprises, glasshouses are confined to a small minority of holdings, about 3 per cent. in England and Wales and 2 per cent. in Scotland, although Merioneth, Montgomery, Radnor and Sutherland are the only counties with none. In England and Wales two-thirds of the holdings have under $\frac{1}{4}$ an acre of glass, the smallest acreage which could support one man, but half the acreage is on the 7 per cent. of holdings with an acre or more; of the 26 growers with 10 or more acres (who account for 13 per cent. of the glass), 11 are in the Lea Valley. About three-quarters of the acreage is heated.

As a commercial enterprise, glasshouse cultivation originated in the Worthing area, the Lea Valley and north-west Kent in the late nineteenth century. Glasshouses around London were gradually displaced as the city grew and new centres arose elsewhere, around Blackpool in the 1920s, following a disaster to local strawberry stocks, and around Hull in the 1930s, largely developed by Dutch immigrants. An expansion after the Second World War, when horticultural produce was in short supply, has been followed by a steady decline since 1951 and the acreage fell 11 per cent. between 1957 and 1967, mainly through the loss of half the acreage in the Lea Valley. The share of the Eastern and South-east Regions declined, and the East Riding was one of the few areas to show a marked increase. Much of this glass is thus obsolete; at the 1966 glasshouse census, 42 per cent. of the acreage in England and Wales and 67 per cent. of that in Scotland had been erected before 1945. However, a new, better-equipped glasshouse industry is emerging, encouraged by generous capital grants under the Horticultural Improvement Scheme; two-thirds of the acreage under glass was automatically heated in 1967 and over half automatically watered, compared with 36 per cent. and 23 per cent. respectively in 1963. There have been considerable changes in the crops grown, with greater emphasis on flowers and pot plants, a result of the relaxation of war-time cropping and fuel restrictions; tomatoes, lettuce and cucumber remain the staple crops and, of these, only cucumbers, with half the acreage in the Lea Valley, exhibit any marked localisation.

Past commitment is thus important in the distribution of cropping under glass; the Lea Valley, where 70 per cent. of the glass was erected pre-war and production is handicapped by labour problems and atmospheric pollution, has already been mentioned. Such inertia is understandable with the high cost per acre, but it may seem strange that, with automatic control of temperature, water and, increasingly, artificial atmospheres, natural conditions should play any part. Yet light is essential for growth and affects yields markedly, and the considerable regional differences in the intensity of winter light cannot be made good by the grower; on average, it is highest in the south-west and lowest in Scotland, but there are also comparatively favourable localities in coastal Lancashire and the East Riding, both centres of glasshouse cultivation. Atmospheric pollution, however, reduces light intensity in the Lea Valley well below the values to be expected from its latitude. The range of winter temperatures is also relevant, for it determines the length of the growing season in cold glasshouses, and also the extra heat necessary to maintain a given temperature. Growing tomatoes under optimum conditions, for example, requires an average night temperature of 60° F. and the difference between this and the mean night temperature in the main growing season from February to July determines how much extra heat should be supplied in any locality. The difference between Eskdalemuir in Dumfriesshire and Worthing is 6·6° F., and L. G. Bennett estimated in 1963 that compensation for this would require an additional expenditure on fuel of some £750 per acre in Scotland; he also estimated the loss of output through differences in light in Sussex and Scotland at some £1,550 per acre, although the Lea Valley was also disadvantaged by its poor light to the extent of £590 per acre, compared with Worthing.

These were estimates computed from climatic data and not recorded results. In practice, heating costs vary considerably with the type of glasshouse and the level of management. R. R. W. Folley and R. A. Giles have assembled data for one year showing that, while average yields decreased northwards, the differences in fuel costs for early crops grown in the Clyde Valley and in south-east England were small, although the Clyde Valley crop was later maturing and the differences increased markedly with later crops. Furthermore, the handicaps of shorter winter days may be offset by longer summer days and by the better quality that may result from slower growth. Folley and Giles also show a northward shift in the occurrence of optimum climatic conditions for tomato-growing as the season advances and this offsets some of the advantages of early crops to growers in the south.

The differences in heating costs may be offset to greater or lesser degree by differences in return arising from location in relation to markets. It is estimated, assuming uniform per capita consumption, that some 70 per cent. of tomatoes are grown within 20 miles of their market. Local tomatoes still command a premium because of freshness and marketing costs are often lower; average monthly prices in the Clyde Valley in 1964 were consistently higher than in south-east England, although the differences ranged from 35 per cent. in May to 7 per cent. in June and were generally about 15 per cent. Thus in no area are the climatic handicaps sufficiently severe to offset the advantages that arise from access to local markets. Similar considerations apply to other glasshouse produce, although such premiums may be affected by the revolution in retailing that is replacing the local retailer by the supermarket chain; fortunately for the small local producer, there seem to be few economies of scale in production under glass. Nevertheless, modern equipment can be very significant: glasshouses developed in the Netherlands in the 1950s permitted a 40 per cent. reduction in labour requirement and the glasshouse industry is the branch of horticulture in which the most rapid advances are being made.

Most of the glasshouse acreage is managed by specialists. It has been estimated that over four-fifths of growers depend entirely on their glasshouses for their livelihood; 84 per cent of glasshouse produce is grown by them and the remainder by market gardeners.

TYPES OF FARMING

The relationships between the different kinds of horticulture and type of farming have already been indicated. According to W. L. Hinton, farmers accounted for some 37 per cent. of total horticultural output in 1965, providing two-thirds of the vegetables, half the field flowers and a quarter of the fruit. Market gardeners accounted for 26 per cent. of the output, but they did not dominate production in any sector, producing half the field flowers, a third of the field vegetables and one-sixth of both fruit and glasshouse produce. The remaining horticultural output was provided by fruit growers, with 16 per cent. of the total output and three-fifths of the fruit crop, and glasshouse specialists, with a fifth of all horticultural output and five-sixths of the produce grown under glass.

Analyses of the regional contribution of horticulture on different types of farms in England and Wales on the basis of standard labour requirements have been published only for commercial orchards. These show that, except in Wales and South-west England, orchards are mainly on horticultural holdings, which have 76 per cent. of the orchard acreage, with proportions ranging from 90 per cent. in the Eastern Region to 63 per cent. in the West Midlands, and that most of the remainder are on mixed or cropping farms. In South-west England, with a tenth of the acreage under commercial orchards, only 39 per cent. of the orchard acreage is on horticultural holdings, 27 per cent. is on dairy holdings and 27 per cent. on mixed holdings, while in Wales, with only 1 per cent. of the acreage, horticultural holdings have only 18 per cent. In England and Wales 76 per cent. of the soft fruit is on horticultural holdings and nearly all the rest on cropping and mixed farms. No doubt an analysis of the vegetable acreage would show an association with cropping and mixed farms, for less than half the acreage of vegetables is grown on horticultural holdings. By contrast, the acreage under glass is almost entirely on horticultural holdings. For fruit and glasshouse produce, therefore, there is much truth in the dictum that 'as an industry horticulture is more diverse than agriculture; as individual producers, growers are often more specialised than most farmers'.

CONCLUSION

This summary of the distribution of horticulture in Great Britain has given a glimpse of a highly complex industry, although some of its branches lie outside the field of farming proper. Horticulture, more than any other enterprise, is increasingly concentrated in those localities where it can most advantageously be practised, for, as the most intensive form of crop production, it can generally outbid all other enterprises (except intensive poultry) for the use of land. By reducing transport costs, improvements in transport have greatly enlarged the area where such crops might be grown, but horticultural crops in general seem to find their greatest advantage in eastern England. The degree of concentration provides one measure of this advantage, but average output per acre of land under horticultural crops provides another, when allowance is made for the regional differences in the composition of horticultural output. Thus, while Eastern England (with most of the vegetables) and South-east England (with most of the fruit) averaged £170 and £190 per acre respectively in 1965–6, the West Midlands and South-west England averaged only £132 and £142 respectively.

In outline, the distribution of horticulture can be plausibly explained, although the reasoning is based on assertions of advantage and on an examination of acreage trends. In detail, it is much more difficult to provide convincing explanations because of the lack of information. The origins of many enterprises seem to have been fortuitous, such as the failure of the strawberry crop which led to the emergence of the west Lancashire glasshouse industry, the migrant from Kent who founded the Fenland fruit industry and the chance introduction of asparagus growing to the Breckland. In part the reason is the complexity of factors, and it must often be difficult for would-be growers to find an adequate basis for the selection of an optimum location; according to R. R. W. Folley and W. L. Hinton, nearly sixty permutations of factors influence growers' decisions about marketing alone. Growers' choice is often affected by local practice and tradition: few of the growers interviewed by L. G. Bennett in west Cornwall could offer any plausible explanation for their choice of crops and few of the answers they gave were economically valid. Although geographical aspects of horticulture are better documented than those of any other enterprise, much further work is needed to explain satisfactorily the distribution of this diverse and complex industry, which has been likened to a stained glass window, in which 'the whole picture is dependent on the formation of all the fragmentary pieces'.

FURTHER READING

L. G. Bennett, *Scale and Intensity in Market Gardening* (1963), Dept. of Agric. Econ., Univ. of Reading

Ibid, *The Diminished Competitive Power of the British Glasshouse Industry through Mal-location*, (1963) Dept. of Agric. Econ., Univ. of Reading

R. R. W. Folley and R. A. Giles, *Locational Advantage in Tomato Growing* (1966)

R. R. W. Folley and W. L. Hinton, *British Fruit Farming*, Rept. No. 64 (1966), Farm. Econ. Branch, Univ. of Cambridge

W. L. Hinton, *Outlook for Horticulture* Occ. Paper No. 12 (1968), Farm Econ. Branch, Univ. of Cambridge

Ministry of Agriculture, *Examination of the Horticultural Industry 1967* (1968)

Ibid, *Horticulture in Britain: Part I, Vegetables* (1967)

CHAPTER 14

The Pattern of Farming

The enterprises discussed in the preceding six chapters do not exist in isolation. They compete for land and other resources and are combined in varying strength on individual farms in various farming systems which also differ in their degree of localisation in different parts of the country. Whatever the relationships of these enterprises with the natural and the man-made environment and with the other factors of production, their regional importance certainly expresses the sum of individual assessments of their comparative advantage made by the occupiers of agricultural holdings in the light of their own circumstances and preferences. This chapter considers how these enterprises may be grouped into types of farms and, more especially, how to measure the relative importance of both enterprises and farming types throughout the country.

The identification of farming types and the assessment of their regional strength involves three separate, though related, procedures: the recognition of individual enterprises; the combination of these by holdings to give types of farms; and the grouping of farms in type-of-farming areas (or, where data for individual holdings are not available, type-of-enterprise areas). This procedure has not always been followed by investigators and sometimes such areas have been mapped without either the individual enterprises or the types of farm being classified, generally because of lack of data or resources, but it represents what is desirable. Since the methodological problems of identifying types of farm resemble those of identifying areas of similar farming they can conveniently be discussed together. The former can, of course, be undertaken independently of the latter; for example, while the Ministry of Agriculture has classified enterprises and farm types on several occasions since 1960 and is now doing so annually, only recently have attempts been made to examine their spatial characteristics.

ENTERPRISES

Reference has been made in previous chapters to types of farming and both they and the individual enterprises have been treated as self-evident, but they are not. The term 'enterprise' can signify a single crop or class of livestock or a group of related activities, which together form a distinct branch of farming, and the choice depends on the purpose of the classification, the size and nature of the area and the complexity of classification that can be tolerated. Wheat, barley and oats could be regarded as separate enterprises and might be treated as such in studying a comparatively small area where farms were similar except for the cereals grown; but they could also be regarded as a single enterprise, cereal growing, as in *The Pattern of Farming in the Eastern Counties*. Potatoes might similarly be treated as one, two or three enterprises, depending on the importance of seed, early and main-crop potatoes, or as part of a composite enterprise of cash crops. Recognising livestock enterprises is generally more difficult, partly because of the variety of purposes for which different classes of stock may be kept, and partly because they depend to varying degrees on other enterprises, as, for example, the growing of fodder crops. Thus, dairy cattle may be regarded as one enterprise (Figure 20) or as two, milk production and the rearing of herd replacements, while beef production might be treated as one enterprise, as two (with fattening considered separately), or as three (breeding, rearing and fattening). Such composite enterprises must be constructed from farm records of some kind, and probably the greater the number of enterprises recognised, the more complicated will be the types of farming and, *a fortiori*, the type of farming areas.

In studying any large area one must depend on the agricultural censuses, although these are not an ideal source of data. Thus, while it may be desirable to recognise cash crops as one enterprise, the censuses provide no information about crop disposal, and, although fairly reliable national estimates can be made about the proportion of a given crop that is sold, how this varies from area to area or from farm to farm is not known. Similar problems arise over the identification of livestock enterprises; for example, dairy followers cannot unambiguously be identified from the census data in England and Wales. Grass and forage crops present other difficulties; treating them as separate enterprises detracts from the importance of the individual livestock enterprises they support, yet there is no satisfactory way of allocating them between the different classes of livestock. Nor can the varying degrees of dependence on purchased feed be determined.

A different problem arises with pigs and poultry. These enterprises require little land and generally have few links with, and little impact on, the other parts of a farming system, more particularly as the scale of individual pig and poultry units grows and industrialised methods become widespread; furthermore, the size of such enterprises can be varied very rapidly, a handicap to the comparison of changes in farming type over time. It could therefore be argued that these enterprises should be omitted from any discussion of farming types, or relegated to a second stage of classification, as has been done in the analysis of types of farming in East Anglia. It could also be argued that, since pigs and poultry are the dominant enterprises on only a relatively small number of specialist holdings and are rarely dominant in any region, little would be lost by excluding them. This argument would have great merit if the basis of classification were land use, but it is not accepted for the subsequent analysis, where the emphasis is on contribution to the farm as a business.

There is no one objective way to identify the individual enterprises, although the Ministry of Agriculture found that most items on the agricultural census form fell naturally into a few enterprises. It recognised five major enterprises (more strictly, groups of enterprises), viz, dairying, livestock (i.e. rearing and fattening beef cattle and sheep), pigs and poultry, cropping, and horticulture (which includes all fruit and vegetables, whether grown on a field scale or not). These can be disaggregated if necessary; for example, in a later stage, involving the sub-division of cropping farms, cereals are separately distinguished from other crops. Grass has been allocated between the grazing livestock on the basis of their requirements in starch equivalent. In Scotland a different approach was adopted, sheep being considered as a separate enterprise and cattle fattening being separated from the earlier stages of beef production. Data for individual farms were not available for this book and enterprises were identified by districts in England and Wales and by parishes in Scotland; but, so far as the census forms allowed, a uniform approach, differing slightly from that used by the two agricultural departments, was adopted throughout Great Britain. Thus, wheat, potatoes, sugar beet, hops and half the barley crop were treated as cash crops, and the remaining crops and the grass acreage allocated to the different classes of livestock in proportion to their relative importance as measured in livestock units. Numbers of young dairy herd replacements in England and Wales were estimated from the ratio of dairy cows to other cows.

With the number of enterprises agreed, the basis of comparison of crops and livestock must be decided. It is obviously meaningless to compare directly, say, acres of grass with numbers of livestock (although they can be related in density of stocking) or numbers of sheep with those of poultry (although these could both be converted to livestock units). Numbers of livestock could be expressed in terms of their land requirements, but, even if differences in land quality and carrying capacity could be satisfactorily allowed for, this approach would exaggerate the importance of extensive enterprises, such as hill sheep rearing, and minimise that of enterprises requiring relatively little land, such as horticulture. A monetary measure, such as sales of farm produce, would be more satisfactory, but is not generally possible because suitable data are lacking. For any large area, the only practical method is the use of standard factors applied to the acreages and numbers of livestock recorded in the agricultural censuses. Two possible sets of factors are standard outputs (expressed in money value) and standard labour requirements (expressed in standard man-hours and man-days), and the latter has been chosen because it is used by the agricultural departments. Their decision rested largely on the greater stability of standard labour requirements, an important consideration, since one objective of their classification of enterprises and farms was to throw light on changes from year to year; for, while standard labour requirements do change with other agricultural developments, such as greater mechanisation, they do not fluctuate from year to year as prices do. In any one year, however, the consequences of using one set of factors rather than the other are small, and, for the Ministry of Agriculture, 'the selection of standard man-days was made largely on grounds of convenience' (C. J. Brown and L. Napolitan).

Standard labour requirements, which can be multiplied by the appropriate crop acreage or number of livestock to convert all the census data to a common basis of standard man-days (s.m.ds), have been devised for each item in the June census. For example, in 1965, the standard for an acre of barley in England and Wales was 2 s.m.ds per acre, for an acre of commercial orchards, 33 s.m.ds, for a dairy cow, 12 s.m.ds, and for a broiler, 0·01 s.m.ds; all the factors are listed in a supplement to the Ministry's publication *Type of Farming Maps of England and Wales*. Of course, it cannot be assumed that exactly the same inputs of labour are required everywhere, especially for crops; barley requires less labour per acre on the large fields of the gently undulating chalklands, where combining is easy, than on the small fields and steep slopes of the upland margins. Attempts to overcome this difficulty raise as many problems as they solve, for using different factors in each region leads to sharp discontinuities at regional boundaries; but the problem is not serious in practice, for crops are generally unimportant in those areas where actual inputs per acre are above average, while the labour requirements of livestock are unlikely to vary widely from place to place.

Once the number of enterprises, their composition and the factors necessary to convert all data to a common basis have been decided, the number of standard man-days devoted to each enterprise can be calculated, a feasible procedure for any large area only with the aid of electronic data-processing equipment. These totals then provide one possible measure of the importance of each enterprise and Figure 20 shows the share of the principal enterprises discussed in preceding chapters.

x

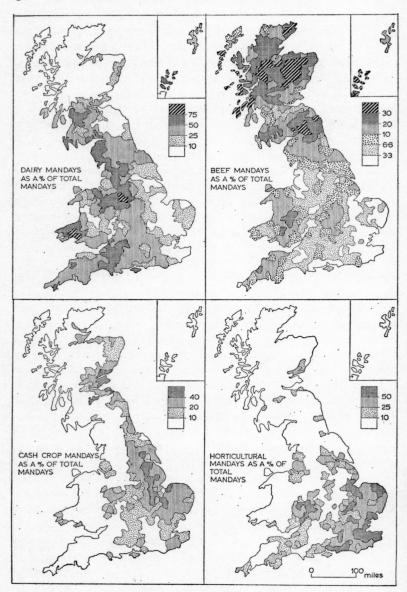

Figure 20. Major enterprises
Source: agricultural censuses 1965

TYPE OF FARM

Identifying types of farm by the combination of enterprises on each holding is possible only where data for individual holdings are available; with summary data, the next step is the identification of areas distinguished by similar combinations of enterprises, although the procedure followed is very like that used for grouping enterprises into farming types and the latter into types of farming area. Where data for individual holdings are available, it has been the practice to confine classification to full-time farms. These have been defined by the total number of man-days or man-hours required by the holding; in England and Wales the threshold between full-time and part-time holdings has been fixed at 275 s.m.ds, while in Scotland the basis changed from 2,400 man-hours in 1947 to 2,000 hours in 1962.

Two approaches to the identification of farm types are possible. Data may be examined subjectively against a general knowledge of farming structure, helped by various guide lines which can be overridden in the light of the classifier's own knowledge and judgment; alternatively, rules can be devised so that each farm can be allocated automatically and unambiguously to a particular type of farm. The former approach was used in the type of farming map of England and Wales prepared by the Ministry of Agriculture in 1939, although this was not strictly a classification of farms, but the recognition of the typical or most common type of farm in each area; the basis of grouping was a subjective appraisal of dominant enterprises, using information on the share of gross and net output, land, labour and other factors, although some account was also taken of secondary enterprises. The classifying of farms in Scotland in 1947 was rather similar; a set of empirical standards left some discretion in the allocation of farms to different types; for example, a hill sheep farm generally had about 95 per cent. of its land in rough grazing and permanent grass, there were normally at least 400 ewes and, although cattle rearing might be undertaken, there were not more than 20 cattle per 100 ewes. The advantage of such an approach is its flexibility but it demands intimate knowledge from those undertaking the classification and it can neither be repeated nor used to establish the nature of changes over time.

In the second approach, devising precise statistical rules, there are also several possibilities. A mainly qualitative classification could be adopted, with enterprises placed in rank order according to the number of s.m.ds required for each enterprise, and the farm would then be classified according to the leading enterprise, as in Figure 21, which shows the leading enterprise in each district. The second and lower ranking enterprises could also be taken into account, although the classification would become progressively more complex. The weakness of this approach is the considerable differences in the number of enterprises on each holding (or in each district or parish) and in the importance of the leading enterprise. The leading enterprise may be the sole enterprise; but it may be only the first of several almost equally important enterprises, although Table 36, showing the number of mixed holdings, i.e. those where no enterprise accounts for more than 50 per cent. of all s.m.ds, indicates that the latter situation cannot be very common. It is more likely to occur in parish or district summaries, since these represent averages for numerous holdings.

Alternatively, a quantitative method, specifying the numerical values of any grouping, could be employed. Here, too, are several possibilities. From frequency analyses of the proportion of s.m.ds attributable to the various enterprises, class intervals or cut-off points could be determined, which would enable holdings to be allocated unambiguously to clearly-defined types. A less rigid approach would require the application of some mathematical operator to the data for each holding, as in John Weaver's crop combinations, in which the actual frequency distribution was compared with some ideal grouping to establish a best fit by the method of least squares.

The Ministry of Agriculture based its classification on the proportion of total man-days attributable to the leading enterprise falling within pre-defined limits, selected after the analysis of a mass of census data. All full-time holdings were grouped according to the main enterprise which accounted for more than half their total man-days, those with no such dominant enterprise being classified as mixed farms; for

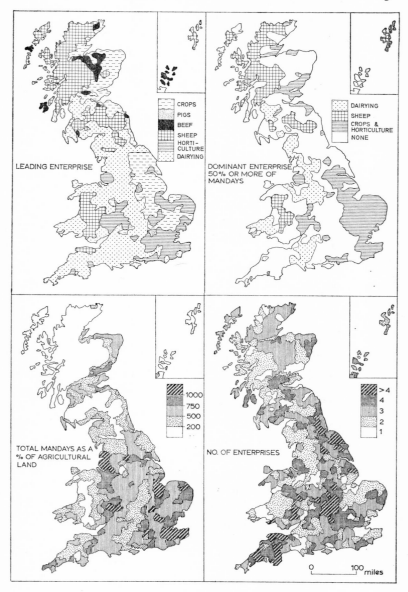

Figure 21. Enterprises and mandays
Source: agricultural censuses 1965

example, a holding with over 50 per cent. of its s.m.ds in dairying was a dairy farm, while one having 45 per cent. in dairying and no other enterprise with 50 per cent. or more was a mixed farm. Three levels of dominance were recognised, 50–75 per cent., 75–87½ per cent., and over 87½ per cent., although subsequently only one type with more than 75 per cent., Predominantly Dairying, was separately distinguished.

TABLE 35
DEGREE OF DOMINANCE OF ENTERPRISE ON HOLDINGS IN ENGLAND AND WALES, 1960

Degree of Dominance (%)	Dairying	Horti-culture	Cropping	Sheep	Beef Cows	Pigs and Poultry
		Percentage of all man-days				
87½–	11	35	9	12	6	11
75–87½	23	14	14	10	7	5
50–75	38	18	29	20	19	6
25–50†	5	5	11	11	13	9
–25†	3	10	17	19	25	33
–50*	14	10	17	20	18	17
Part-time holdings	6	8	3	8	12	18
ALL HOLDINGS	100	100	100	100	100	100

* As subsidiary to a major enterprise (50%)
† No enterprise dominant
Source: L. Napolitan and C. J. Brown

These major groupings were then subdivided. Livestock farms were divided into three categories: mostly sheep farms, with sheep accounting for 75 per cent. or more of the man-days attributable to beef cattle and sheep; mostly cattle, with 75 per cent. attributable to beef cattle; and mixed cattle and sheep, the remaining livestock farms. Mostly pigs, mostly poultry, and pig and poultry holdings were similarly identified and cropping farms sub-divided, although only two, mostly cereals and general cropping, remain of the four categories originally identified. By contrast, horticulture, originally classified as one type, has been divided into three, mostly fruit, mostly vegetables and general horticulture. The subsequent discussion will focus attention on these main categories. Table 35 records the breakdown of the different enterprises, according to degree of dominance. It shows a progression from dairying, with 72 per cent. of its man-days on holdings with dairying as the main enterprise, through horticulture, cropping and livestock to pigs and poultry, where the proportion is only 22 per cent. even with these two enterprises considered together. There is also a corresponding increase in the proportion of subsidiary enterprises and a tendency, absent only in cropping farms, for the proportion attributable to part-time holdings to increase.

TABLE 36
NUMBERS OF HOLDINGS OF EACH TYPE IN 1965

ENGLAND & WALES (000s)

Dairying	Livestock	Cropping	Pigs and poultry	Horticulture	Mixed
60·8	24·7	26·6	9·6	15·0	19·5

SCOTLAND (000s)

Dairying	Arable	Rearing with Arable	Rearing with intensive livestock	Arable Rearing and Feeding	Cropping	Hill Sheep	Upland	Intensive
7·0	4·8		1·1	2·0	3·5	1·3	2·9	1·2

Source: Agricultural Departments

In Scotland, both the classification and the method of classification are rather different, although there is a broad equivalence between the different types (Table 36). Subdivisions of farming types in Scotland reflect the greater importance of beef cattle and sheep. Eight types of full-time farm were recognised in 1962, three of which are mixed types (rearing with arable, rearing with intensive livestock and arable rearing and feeding); three of the main types were subdivided to give fourteen types in all, the same number as in England and Wales. The method of classifying farms, was, however, more complex and some holdings could be classified in several ways, so that a definite sequence had to be followed; the approach resembled that used in classifying the 1947 data in that empirical rules were devised, specifying appropriate ratios of different categories of stock and crops. For example, a dairy farm was a milk-selling farm registered with a milk marketing board and with dairy cows accounting for at least 25 per cent. of all man-days (except pig and poultry s.m.ds.), but excluding two categories, those with very small herds and those with a high proportion of other intensive enterprises; dairy farms were then divided into dairy with sheep farms (those with at least 20 per cent. of man-days in sheep), and farms with low or high self-sufficiency. The other categories in order of classification were: intensive farms, subdivided into horticultural, poultry and pig farms, each with 60 per cent. or more of their standard man-days in those enterprises, and mixed intensive; hill sheep farms (subdivided according to the ratio of cattle and sheep); cropping farms; upland farms; and rearing with intensive livestock, arable rearing and feeding, and rearing with arable farms, the last a residual type which included all farms not allocated into any preceding category. For Figure 21, all hill sheep farms, upland farms and livestock with arable or intensive livestock farms were classed as livestock farms.

These different approaches illustrate the problems of classification. The system used in England and Wales is simple and the process of classification easy, but some categories probably include too wide a range of types, e.g. dairying. The Scottish approach is more flexible, but also more complex and less easily applied, and appears to have been tailored to fit known types of farm. Whether, given the somewhat arbitrary dividing line between full-time and part-time holdings, the exclusion of part-time holdings from both classifications is justified is debatable, especially in north-west Scotland where part-time holdings are numerous. A classification of type of farming areas which included part-time holdings as one type of farm revealed that such holdings were the leading type over extensive areas of north-west Scotland.

TYPE OF FARMING & TYPE OF ENTERPRISE AREAS

Once types of farm have been established, the classification of type-of-farming areas or, where farm data are not available, type-of-enterprise areas, may be attempted. Similar procedures can be used to those employed in the classification of types of farms, although this double classification may lead to the dominant enterprise not being that of the dominant type of farming, as where 51 per cent. of the farms are classified as dairy farms, although they have only 51 per cent. of the man-days devoted to dairying. Although this extreme situation is unlikely, individual enterprises may nevertheless be less important in an area than the type-of-farming classification suggests.

There is, however, one problem peculiar to the identification of the type of farming or enterprise areas. A farm is a unit with identifiable boundaries and the enterprises undertaken within these boundaries define the farming types, however these are selected; similarly, where only summary data are available, the parishes, districts or counties have finite boundaries which circumscribe the data. In identifying type of farming areas, no such areal frame is given, and there are several possibilities. One solution is to avoid the problem by allowing the facts to speak for themselves by merely plotting the individual types of farming on a composite map, as in the type of farming map which forms the frontispiece to *The Pattern of Farming in the Eastern Counties*. A second solution is to plot the distribution of farms and attempt to identify type of farming areas subjectively by inspection; this approach was used in compiling an unpublished type of farming map of Scotland, using 1956 data, and also for the type of farming map in the Ordnance Survey's ten-mile series, although here the individual farms were not all plotted. The principal danger of this approach is that boundaries will be drawn, consciously or unconsciously, to coincide with physical features. A marked change in physiography is indeed often accompanied by a change in farming type, as a transect from the Chiltern scarp to the Vale of Aylesbury shows; but such a relationship should not be assumed in the absence of farm boundary maps, and it is not uncommon for farms in such an area to be long and narrow, and so contain contrasting enterprises on different types of land. Where the farm can be identified only by reference to some administrative area, e.g. a parish, it is unwise, in the absence of any supporting evidence, to assume such a relationship, e.g. that all dairy farms are on low-lying grasslands.

Another solution is to use some pre-determined areal unit. Obvious examples are parishes (necessary where, for reasons of confidentiality, this provides the only locational reference) or squares of the National Grid. The latter have been used by the Ministry of Agriculture in their publication, *Types of Farming Maps of England and Wales*, and a similar approach was used by R. Bennett-Jones in his examination of types of farming in the East Midlands, although, for unexplained reasons, he adopted a rectangular grid. The only valid objection to such an approach is that such units almost certainly, by definition, cut across farm boundaries and any grouping of farms which exists; but it is more likely to be opposed because of its 'unnatural' appearance. A regular grid has one great merit over administrative areas: it is uniform in size and shape. Whichever unit is chosen, such an approach produces not a map of agricultural regions, but one of areas within which particular types of farms predominate or are found in specified proportions. The degree of fragmentation of pattern depends partly on the complexity of farming and partly on the size of the areal unit; other things being equal, small areas, which are more likely to reflect the influence of large farms or random variations in farming patterns, will probably show a more varied pattern than large.

The last approach avoids the choice of areas and boundaries by having none. Farms are treated as point samples and isopleths interpolated between them, either by inspection or, preferably, by means of some contouring algorithm such as is incorporated in automatic mapping routines, which always produces the same results from the same data. Alternatively, as J. Tarrant argues, the results may be interpreted by the identification of trend surfaces, which are both a substitute for regions and an interpretation of any regional tendencies the distribution of such farms may exhibit. The weakness of this approach is that only two classes of farm may be considered at any one time.

The balance of advantage appears to lie in using some pre-determined area, especially if this is fairly small in relation to the size of the area under consideration; for the larger the unit area, the greater the generalisation that results. This is illustrated by comparing the maps in Figure 22, the one based on data for administrative counties, ranging in size from 141 km^2 to 109,000 km^2, the other based on data for

THE PATTERN OF FARMING 317

10 km squares (100 km²,) of the national grid; although the two maps are not strictly comparable, the contrasts arising from differences in the degree of generalisation are clear.

Once the areal unit is chosen, two other decisions must be made, the basis of comparison and the range of permissible variation within each unit. There are at least three criteria for

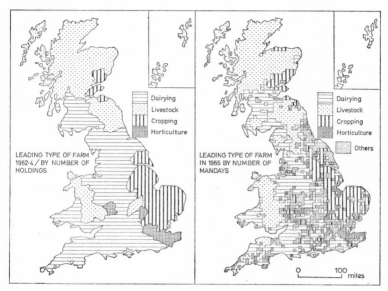

Figure 22. Types of farming
Sources: agricultural censuses 1965 and adaptation of a map by B. M. Church et. al.

deciding the importance of different types of farm: number of holdings, acreage and total man-days. The number of holdings of different types is sometimes the only information available and the proportions are simple to calculate. The main disadvantage is that farms vary greatly in size (whether measured by acreage, standard output or standard labour requirements) and there are differences in the size distribution among the various types of farm. The choice between acreage and standard labour requirements depends largely on the purposes of the classification; one by acreage emphasizes the amount of land in each type of farm, irrespective of the intensiveness of its enterprises. A hill sheep farm, of large area, but representing a small business, carries more weight than an intensive horticultural holding occupying only a small acreage. The reverse is true on a basis of standard man-days, and an area such as south-east Hertfordshire might be classified as predominantly horticultural because of the high rating given to glasshouses, although they occupy little of the area. One approach is no more correct than the other, since they serve different purposes; and in practice the differences between them are much smaller than might be expected, for the intensive types are often highly localised while the more extensive types of farming are largely confined to the uplands and the upland margins where few other enterprises are possible. Thus, with allowance for the small scale of the maps and for crudeness of the county as a statistical unit, the maps resulting from the different approaches used in the construction of Figure 22 are strikingly similar.

A similar approach is possible in mapping enterprises by area, but here, even more than with type of farming, reference only to the leading enterprise can give a wrong impression; attempts to incorporate information on the range of enterprises in a classification of areas have not been very successful, the resulting maps being too complex to be easily comprehended. It is probably better to examine the individual enterprises separately, firstly for degree of dominance, then for subsidiary enterprises, and finally for the number of enterprises. Simply counting the latter in each administrative area may be misleading in that most enterprises are on some farms in nearly all areas even if the acreages or number of man-days devoted to them are small; using some statistical procedure which eliminates from consideration enterprises of little significance (as in Weaver's Crop Combinations) or fixing arbitrary minimum values is preferable. The former approach has been used for Figure 21 in computing the numbers of enterprises recorded in each district or parish; these refer to the number included in the enterprise combination calculated for each administrative area by a modified version of Weaver's formula. The range of enterprises varies with the size of holding, the size of the area under consideration and its physical character; the larger the holding, the more likely it is to have a larger number of enterprises, a fact suggested by the tendency for 'predominant' holdings, i.e. those where the leading enterprise accounts for over 75 per cent., to be more common among the smaller holdings; similarly, the larger the area, the greater the variety of enterprises it is likely to contain. No very clear picture emerges from Figure 21, although the range is least in the uplands and in eastern counties, and greater in the south and Midlands than in the north.

REGIONAL TYPES OF FARMING

The distribution of each enterprise and its relationship to types of farm, have already been described and this section will provide only a comparative summary. Table 37 indicates the importance of each enterprise on the different types of farm. Dairy cattle show the highest specialisation of the enterprises listed, with 75 per cent. of all dairy cows in England and Wales and 89 per cent. in Scotland on dairy farms. No other enterprise is so concentrated on any one type of farm, although in all categories except breeding pigs in England and Wales and laying fowls in Scotland each enterprise is most prominent on the type of farm which bears its name. Except for intensive holdings in England and Wales and intensive holdings and mainly livestock types in Scotland, the range of enterprises on other types of holding is surprisingly wide.

Numbers of full-time holdings of each type are given in Table 36. Because of differences in classification, Scotland is treated separately from England and Wales. The pattern of farming in Scotland is simpler than that south of the Border, although this is obscured in Table 38 by the larger number of farming types, especially the sub-divisions of livestock categories. This table provides a summary of the regional importance of the different types. The first set of figures records the percentage distribution of the various categories of full-time holdings in each region and confirms the prominence of rearing farms in the North-east, of cropping farms in the East-central Region and of dairy farms in the South-west. The remaining figures show the proportion of all full-time holdings by number, acreage, and standard man-days; in each region the main contrast is between the distribution by acreage and that by standard man-days, the former exaggerating the importance of the livestock types and minimising that of intensive enterprises, while the reverse is true in the classification by man-days. Areally, the most specialised region is the Highlands, with 67 per cent. of its acreage in hill sheep farms; with the various rearing types amalgamated, the North-east ranks second. By

TABLE 37
PERCENTAGE DISTRIBUTION OF THE MAIN ENTERPRISES BY TYPE OF FARM, 1965

Type of Farm	Dairy Cows	Beef Cows	Breeding Sheep	Breeding Pigs	Laying Fowl	Wheat	Barley	Potatoes
ENGLAND & WALES								
Dairy	75	5	13	19	15	14	17	7
Livestock	2	55	55	5	3	6	7	2
Pigs and Poultry	1	1	1	19	54	2	3	1
Cropping	3	13	8	18	4	55	48	66
Horticulture	—	1	1	4	2	4	2	9
Mixed	13	13	14	19	12	16	18	11
Part-time	6	12	8	16	10	3	5	4
ALL HOLDINGS	100	100	100	100	100	100	100	100
SCOTLAND								
Hill sheep	1	7	31	—	1	—	—	—
Upland	2	33	23	2	6	2	3	2
Rearing with arable	2	29	14	10	14	8	15	9
Rearing with livestock	—	3	1	15	12	1	3	1
Arable rearing and feeding	1	5	3	6	6	8	12	8
Cropping	2	8	6	20	10	63	43	53
Dairy	89	6	10	14	20	15	20	20
Intensive	1	—	—	25	14	2	2	2
Part-time	2	9	12	8	17	1	2	5
ALL HOLDINGS	100	100	100	100	100	100	100	100

Source: Agricultural Departments

numbers of holdings and standard man-days, however, the South-west is the most specialised region, with dairy farms accounting for 69 per cent. of the standard man-days and 60 per cent. of the full-time holdings. The distribution of farm types in Scotland is, of course, greatly influenced by the large tracts of rough grazing, occupying two-thirds of the country and largely devoted to hill sheep or upland farms; but in the peripheral lowlands types of farm are highly localised.

In England and Wales the advantages to be expected from a simpler classification of the main types of farm are largely offset by the heterogeneous character of some of the regions for which data are published, notably the Yorkshire and Lancashire and Northern Regions; but the pattern of farming is, in any case, more complex. Table 39 provides similar information to that given for Scotland, although data on the acreage in each region occupied by holdings of different types are not available

TABLE 38
TYPES OF FARMING IN SCOTLAND, 1962

Region	Hill Sheep	Upland	Rearing with arable	Rearing (intensive livestock)	Arable rearing and feeding	Cropping	Dairy	Horti-culture	Pigs and Poultry
Percentage of full-time holdings of each type, by regions									
Highland	41	22	9	2	8	3	6	1	2
North-east	2	23	65	81	70	25	11	4	20
East-central	13	11	10	4	10	50	9	37	18
South-east	13	9	9	3	8	17	6	17	20
South-west	32	35	7	9	4	5	68	41	41
Scotland	100	100	100	100	100	100	100	100	100
Percentage of full-time holdings in each region, by type									
Highland	21	30	19	1	7	4	16	—	—
North-east	—	9	38	14	17	10	8	—	1
East-central	4	9	12	2	5	43	14	7	3
South-east	7	13	19	2	7	24	17	6	5
South-west	5	16	5	2	1	2	60	4	3
Percentage of acreage on each type of holding									
Highland	67	24	4	1	1	—	3	—	—
North-east	6	17	39	6	12	9	8	—	—
East-central	37	20	11	—	3	21	7	1	—
South-east	29	21	20	1	4	16	8	—	—
South-west	33	22	3	—	1	1	39	—	—
Percentage of standard man-days on each type of holding									
Highland	22	24	16	1	8	7	19	—	—
North-east	—	8	34	11	16	13	16	1	1
East-central	3	6	9	1	5	47	18	9	1
South-east	5	11	20	1	6	30	18	5	1
South-west	4	11	3	1	1	2	69	5	1

Source: Department of Agriculture

in England and Wales. Cropping is the most localised of the major types, with 58 per cent. in the Eastern and East Midlands Regions, followed by livestock rearing and fattening and by horticulture. The distribution of holdings by numbers is very similar to that by man-days. The most specialised region is again one depending primarily on dairy farms, for these accounted for 57 per cent. of all full-time holdings in the South-west; but the most marked regional contrast is between Wales and the Eastern Region, where the two principal types account respectively for 90 and 75 per cent. of all full-time holdings by number and for 83 and 77 per cent. by standard man-days. Table 39 also provides additional information not available in

TABLE 39
TYPES OF FARMING IN ENGLAND AND WALES, 1966

Type of Farm	Eastern	South-eastern	East Midland	West Midland	South-western	Yorks and Lancs	Northern	Wales
Percentage of full-time holdings of each type in England and Wales by regions								
Dairy	3	9	8	16	26	12	10	16
Livestock	1	6	7	12	17	6	17	34
Pigs and Poultry	21	18	9	11	14	19	5	3
Cropping	40	9	18	7	4	14	7	1
Horticulture	35	23	5	14	10	8	3	1
Mixed	9	9	10	13	25	11	13	9
Percentage of full-time holdings in each region, by type								
Dairy	8	33	34	50	57	43	38	48
Livestock	2	10	12	15	15	9	29	42
Pigs and Poultry	9	10	5	5	4	10	3	1
Cropping	50	15	33	9	4	21	13	1
Horticulture	25	23	5	11	6	7	3	1
Mixed	7	9	11	10	14	10	14	7
ALL TYPES	100	100	100	100	100	100	100	100
Percentage of standard man-days in each region, by type								
Dairy	6	24	26	42	52	35	35	40
Livestock	1	5	9	10	12	7	28	43
Pigs and Poultry	9	7	7	5	6	10	4	2
Cropping	46	20	41	15	5	25	14	2
Horticulture	31	32	7	15	8	12	5	5
Mixed	7	11	10	13	17	11	15	8
ALL TYPES	100	100	100	100	100	100	100	100
Percentage of standard man-days on holdings of 1,200 s.m.ds or over								
Dairy	62	55	34	40	36	25	33	14
Livestock	42	31	32	25	24	30	40	24
Pigs and Poultry	71	60	68	58	63	49	59	54
Cropping	69	81	69	79	61	55	48	61
Horticulture	78	82	71	70	63	71	72	89
Mixed	59	75	80	53	55	51	41	32

Source: Ministry of Agriculture

Scotland, recording by type of holding the proportion of all man-days on full-time holdings which are large businesses, i.e., holdings with an annual labour requirement of 1,200 s.m.ds or more. In all regions, such large businesses are most characteristic of horticultural holdings and least characteristic of livestock holdings, irrespective of the regional importance of those types. There is also a tendency, most marked on dairy and mixed farms, for the proportion of large businesses to decline from east to west.

Figure 22, which should be read with Figure 21, provides a synoptic view of the distribution and relative importance of different types of farms. It indicates the differences in complexity of pattern arising from differences of approach and generalisation. It confirms the relative simplicity of farming distributions in Scotland and suggests that in England and Wales regions of relatively simple farming structure alternate with those of more varied farming. The three major dairying regions centred on Cheshire, Somerset and Carmarthan stand out clearly, while eastern England, from north Essex to the Yorkshire Wolds, is a fairly homogeneous region of arable farming. The major uplands also have a relatively simple farming structure with livestock farming predominating, and Kent, together with other localised horticultural regions, is also identifiable as a region of specialised farming. By contrast, most of the Midlands, south and south-east England, north-east England and the south-west peninsula, show great variety of both enterprises (Figure 21) and farming types (Figure 22).

CONCLUSION

This chapter has been concerned primarily with methodology and many of the questions which would otherwise have been discussed have already been anticipated in earlier chapters. The view of British agriculture presented in this chapter, and to a lesser extent through the book, is a static one, not from choice, but because of limitations of space and resources; yet, as E. A. Ackerman has noted, 'all significant areal differentiations have a time dimension', and British agriculture is experiencing rapid changes which will probably strengthen rather than weaken the already marked regionalisation that is characteristic of most areas.

FURTHER READING

D. K. Britton and K. A. Ingersent, 'Trends in concentration in British agriculture', *J. Agric. Econ.* 16 (1963–4)

M. Chisholm, 'Problems in the classification and use of farming type regions', *Trans. Inst. Brit. Geogr.* 35 (1964)

B. J. Church, et. al., 'A type of farming map based on agricultural census data', *Outlook on Agriculture* 5(5) (1968)

J. T. Coppock, 'Crop, livestock and enterprise combinations in England and Wales', *Econ. Geogr.* 40 (1964)

L. Napolitan and C. J. Brown, 'A type of farming classification of agricultural holdings in England and Wales according to enterprise patterns', *J. Agric. Econ.* 15 (1962–3)

P. M. Scola, 'An economic classification of Scottish farms based on the June census, 1962', *Scot. Agric. Econ.* 15 (1965)

Ministry of Agriculture, *Type of Farming Maps in England and Wales* (1969)

Postscript

The agricultural geography of Great Britain described in this book has already been somewhat modified since the text was written. In part, these changes represent a continuation of these already noted; for example, the average size of dairy herds in Scotland has continued to increase, rising from 45 cows in 1965 to 55 in 1969, and the contribution of Ayrshires has fallen further from 78 to 69 per cent. Others are due to hazards such as the foot and mouth epidemic of 1968, which led to heavy slaughter of stock in the West Midlands, or the long cold spring of 1969, which left a legacy of reduced numbers of sheep in north-east England. Further considerable changes are likely over the next decade, although their precise nature will depend on the fate of the United Kingdom's application to join the European Economic Community, on changes in living standards and consumer preferences, on government policy in respect of agricultural prices and taxation, and on the rate of technological change. Will liquid milk, for example, cease to be a major farm product, as suggested in the 1970 Oxford Farming Conference, and will British sheep farmers, admitted to an expanded Common Market in which the United Kingdom is the largest producer and consumer of lamb and mutton, be protected by tariffs against New Zealand producers of fat lamb? Many of these changes, too, will represent a continuation of existing trends: the enlargement of both farms and individual enterprises, declining inputs of labour, greater specialisation by farm and by area, and more intensive production, giving higher yields and stocking rates, are all likely to be features of the 1970s.

The approach to agricultural geography will also change because of the greatly improved facilities for handling data which both government departments and universities will enjoy and because of changes in the outlook of geographers. The energies of geographers have often been exhausted in collecting and reconstructing the spatial characteristics of their data long before their studies were logically complete and there was little rigorous analysis of the features so described. The electronic digital computer has transformed the handling of data and it is now perfectly possible to prepare maps of agricultural distributions by line printer soon after the census data are available in machine-readable form. Geographers will hope that the government agricultural departments will produce routine cartographic analyses showing the location and scale of inter-censal changes, just as they now publish annual volumes of county agricultural statistics and have parish statistics available for consulation. Although it owes little, if anything, to the availability of such equipment, the publication of the volume, *A Century of Agricultural Statistics in Great Britain*, containing a wealth of maps of agricultural distributions, is a welcome first step. Such developments will release energies and resources for investigating aspects which have perforce been neglected, such as the invisible functional or tenurial links between parcels of land and the complex relationships between farm and market. The development of new approaches will also require a better understanding of processes, particularly of decision-making in agriculture and of the factors influencing a farmer's choice of enterprises.

Powerful computational aids will facilitate not only the descriptions of complex agricultural distributions but also their analysis. Statistical techniques of multi-variate analysis are already employed in other branches of geography and they are now being adopted in agricultural geography. These developments are also symptomatic of the changing attitudes of geographers towards their subject. Such changes are common to all the social sciences since they are concerned with highly complex phenomena that depend on human thoughts and actions and so do not lend themselves to experimentation. They have occurred rather later in geography, perhaps because of the difficulty of handling spatial variation and of the predisposition of earlier generations of geographers to give pride of place to physical factors (although the Davisian cycle of erosion provides one of the earliest explicit attempts by geographers to construct a model of the real world, and one which, whatever its limitations, undoubtedly contributed to the rapid expansion of geomorphology). It is a valid criticism that in human geography, not least in agricultural geography, geographers have been too little concerned to provide a conceptual framework for their enquiries or to think in terms of problems that require investigation. Rather they have attempted to study complex situations in their entirety, an approach both praiseworthy and of considerable educational merit, but increasingly impracticable as the information explosion accelerates. It will always be necessary for a geographer to be well informed about the geographical realities of any problem with which he is concerned, and it is appropriate to recall that von Thünen, the father of agricultural location theory, was a land-owner and practising farmer and that his own experience informed his attempts at model building. Nevertheless, a purely empirical approach offers little prospect of any major intellectual development in the subject.

Future agricultural geographers will thus increasingly attempt to understand reality by devising models of ideal situations or by constructing hypotheses to be tested against reality. Given the complexity of British agricultural geography, such attempts will probably be at first concerned to examine manageable aspects, as with location theory. A complete systems approach is unlikely to be very rewarding at this stage, although it will be useful in directing attention to the concepts of structure, function, equilibrium and change. Such model-building has been very little used in agricultural geography, as J. D. Henshall's review has shown, and little progress has been made since von Thünen's attempts to conceptualise the spatial arrangement of agriculture almost a century and a half ago. Even if no results of great practical significance emerge, a more adequate theoretical framework will help both to identify and to make explicit the more important problems needing solution; similarly it will help the teaching of agricultural geography, both by relieving teachers of the impossible task of trying to keep abreast of a rapidly changing world, and by providing students with an explicit structure for their studies.

Increasing collaboration between agricultural geographers and those working in cognate disciplines should also develop; for research into many of the most important topics must be interdisciplinary. Co-operation is desirable between geographers, with their strongly developed spatial sense and their increasing ability to handle spatial data, and economists, whose training encourages both a facility in abstraction and a rigour in analysis that geographers have generally lacked; the study by R. H. Best and R. Gasson of *The Changing Location of Intensive Crops* is an excellent example of such collaboration in the agricultural field. There are equally important frontiers with sociology, psychology and other disciplines to be explored. Whatever the merits of normative economic theory, it will rarely explain satisfactorily the spatial variations in the agriculture of a country. Geographers have begun to explore the possibility of substituting assumptions of 'satisfaction' for those of profit maximisation and to examine the important field of farmers' perception of their own environments and of the resources of their farms, notably of environmental hazards, such as drought and flooding. Geography is in its infancy as a behavioural science, although differences in attitude, often culturally derived, have long been recognised as contributing greatly to the nature of agricultural activity, especially in developing countries.

Model building and hypothesis testing will also require the more explicit evaluation of data. Here, too, technological developments will improve both the quality and the quantity of data, as with the use of false colour and infra-red aerial photography, although such data will be mainly of value for the study of land use in the absence of regularly up-dated cadastral maps. R. J. C. Munton has rightly argued that the farm should be the basic unit of agricultural geography, and clearly many problems will require farm data for their satisfactory solution. This requirement poses other problems of data acquisition, for the number of farms that can be studied at first hand is small; while sampling will help, the sample will have to be large if areal characteristics are to be adequately represented, and the large area occupied by each unit makes the problem of non-response more serious than in most sampling. Farmers may, however, respond more favourably to questions that are directed to solving a clearly-stated problem. In any case, given the large number of factors involved, preliminary analysis of aggregated data should help both in drawing an appropriate sample and in constructing plausible hypotheses.

Some indication of the changing nature of geographical inquiry within the field of agricultural geography is given by the titles of papers read at a seminar held by the Agricultural Geography Study Group of the Institute of British Geographers in January 1969, all but one of them concerned with some aspect of the agricultural geography of the British Isles: they fell into two groups, half concerned with neglected topics and half with new methods of analysis.

(a) The work of the Agricultural Land Service, V. Martin
The inquiry into statutory smallholdings, 1964–1967, M. J. Wise
The labour input in British agriculture, J. Salt
The study of livestock movements—the Irish cattle industry, D. Gillmor
The influence on British agriculture of post-war government policies, I. R. Bowler

(b) A multi-variate study of the agricultural structure of the High Weald, 1887–1953, B. M. Short
Variation in cereal acreage in Great Britain, J. T. Coppock and B. S. Duffield
Trend-surface analysis: an alternative to agricultural regions, J. R. Tarrant
Internal spatial organisation of agriculture in some north Indian villages, P. M. Blaikie
A simulation model for experimentation: control strategies for foot and mouth epidemics, R. Tinline
Farm systems—problems of analysis, R. J. C. Munton.

Such approaches hardly affect the treatment of agriculture in this book, which may therefore attract a comment similar to that passed on a volume of statistical studies by geographers: 'Very interesting. Perhaps someone will now go and write a real agricultural geography'. Before he does so, however, much preliminary work will be necessary; in the meanwhile, this summary volume may suggest topics which can usefully be explored by geographers with an interest in agriculture.

FURTHER READING

C. Clark, 'How I would plan British farming', *J. Farmers Club* (1962)

B. H. Davey, *Trends in Agriculture* (1967), Paper No. 3, Agric. Adjustment Unit, Univ. of Newcastle.

J. D. Henshall, 'Models in agricultural geography' in R. J. Chorley and P. J. Haggett (eds.) *Models in Geography* (1968)

J. Wolpert, 'The decision progress in its spatial context' *Ann. As. Am. Geogr.*, 54 (1964)

Index

The names of individual counties, regions and countries, and occasionally towns, are used liberally throughout this book as aids to description (e.g. 'from Dorset to the North Riding') and no specific importance attaches to them as places. Little benefit would be gained by indexing them and only places to which specific information relates are included in the index (e.g., 'in the Clyde valley, which has half Scotland's 300 acres of glass'). All proper names other than place names are included.

Aberdeenshire, 18
accommodation fields, 56
accumulated temperature, 34
Ackerman, E. A., 323
advisory services, 26, 294
agricultural geography, definition, 1, 2, 8; future methods and studies, 325–30
agricultural land; area, 135; changes, 145–8; composition, 135; losses, 13
Agricultural Land Service, 52, 77
Agricultural Marketing Acts, 22, 27
agricultural policy, 11–28; consequences, 23, 28
agricultural population, composition, 82; density, 81; proportion, 12, 81
Agriculture Act 1937, 23; 1947, 24, 27; 1957, 25; 1958, 27; 1967, 66, 77
Agriculture, Minister of, 54
air pollution, 33, 40, 298
Allan, G. R., 134
Allanson, G., 214
Allscott, 262
altitude, effects, 43; pigs, 229
American Civil War, 14
Ansell, D. J., 106
apples, 289–94; cider, 292; cooking, 291; dessert, 291
Argentina, 17, 179, 201
Argyll, 129, 210
artificial insemination; beef breeds, 183; dairy breeds, 157; numbers of cows, 184
Ashton, J., 86, 106

asparagus, 275, 285
aspect, 45
Astor, Vis. and B. S. Rowntree, 4, 28
attitudes of farmers, 4; to cereals, 249, 259
Attwood, E. A., 221
Australia, 16, 17, 201

bacon; bacon pigs, 22, 224; factories, 126, 230; imports, 21, 225
Balfour Report on Hill Sheep Farming in Scotland, 210
bare fallow, 268
barley, 253–4; cattle fodder, 257; continuous cropping, 259; distribution, 254, 257, 258, 260; exports, 253; fodder, 253; guaranteed prices, 22, 259; human consumption, 254; imports, 253; increase, 260; malting, 253; physical requirements, 254; pig feed, 230, 257; price incentives, 259; susceptibility to disease, 258; varieties, 256, 258; yields, higher, 256
Barnes, F. A., 178
beans; dwarf, 277, 282; fodder, 268; runner, 288
beef, consumption, 179; imports, 21, 179
beef cattle, 179–99; age of slaughter, 182, 197; breeding, 195; contribution, 179, 193; dairying links, 180; data, 179; distribution, 181, 195; England and Wales, 193–5; English calves,

beef cattle—*contd.*
195; enterprises, 304; fat, 22, 181–2; government encouragement, 179, 183; Irish cattle, 195, 197; markets, 124–7; motives, 180; movements, 181–2; output, 195; profitability of stores, 182; regional specialisation, 181; Scotland, 195–9; store, 124, 182, 197; surplus resources, 180; systems of feeding, 198; tradition, 180; type of farm, 193

beef cattle: breeding and rearing, 183–90; beef breeds, 185, 186; beef cows, distribution, 185, 186; calves from dairy herds, 184; calves' movements, 189; calves retained on dairy farms, 184–5; hill cows, breeds, 188; hill cows, environment, 189; suitability of dairy breeds, 183; type of farming, 193

beef cattle: fattening, 190–2; grass, 190; intensive, 192; systems, 190; type of farming, 193; yarded, 192

Bellis, D. B., 244
Bennett-Jones, R., 316
Bennett, L. G., 111, 117, 119, 120, 298, 302
Best, R. H., 328
Bibby, J. S., 53
Biggleswade, 279
Blackpool, 297
Blaikie, P. M., 329
Bowman, J. C., 244
Bowler, I. R., 329
bread; consumption, 14; home-grown wheat, 249, 258
break crop; cereals, 258; field beans, 269; vegetables, 288
Breckland, 275
Brecon, 210
Bristol, 110
British Association, 8
British Egg Marketing Board, 24, 131–2, 242
British Sugar Corporation, 113, 122, 248, 261
British Sugar (Subsidy) Act 1926, 20
Britton, D. K., 7, 107, 114, 134, 273, 323
Britton report, 259

broilers, 234, 239, 243–4
Brown, C. J., 324
Brussels sprouts, 279, 286; frost beneficial, 37, 285
Buchanan, R. O., v, 10
Buccleuch, Duke of, 71
Buckinghamshire, 57, 85
buildings, *see also* housing, farm buildings, 55, 61, 73–5, 163; glasshouses, 33, 41, 279
bulbs, 282
business, size of, 63–5; hill sheep farm, 63; horticulture, 63; pigs, 231; poultry, 236
Butler, J. B., 4, 10, 87
butter, consumption, 14; farmhouse, 22, 164–5; imports, 16, 17

cabbages, 284; as stock feed, 286
Caernarvonshire, 75, 85
Caird, J., 136
calf subsidy, 27, 183
canning; increased consumption, 14; horticultural crops, 275, 277, 286; potatoes, 267; soft fruit, 295
capital investment; cereal crops, 249; egg production, 239; farm buildings, 74; glasshouses, 296; horticulture, 278; orchards, 291; pigs, 232; poultry meat, 243; sheep, 205–6
carrots, 284, 289; soils, 285; for freezing, 286
cash crops; barley, 259, 269; census data, 304; distribution, 138; enterprises, 306; hops, 267, 269; horticultural, 274; potatoes, 263, 269; roots 260; sugar beet, 261, 269; wheat, 249, 269
cattle, *see:* beef cattle and dairy cattle
cattle housing, 41, 163
cauliflower; frost sensitive, 285; summer, 284; winter, 37
celery, 279, 289
census, agricultural, completeness, 57
census, population 1966, 81
Central Council for Agricultural and Horticultural Co-operation, 112
Central Office of Information, 10
cereal marketing; Britton report,

cereal marketing—*contd.*
114; flexibility, 116; merchants, 114; storage, 114; transport, 115–6
cereals, 248–60, and *see* individual crops; acreage per grower, 249; attitudes of farmers, 249, 259; changes between cereals, 248, 256–60; changes in proportion of tillage, 248; characteristics, 248; comparative advantage, 259; economies of scale, 249; enterprise, 304; imports, 256; labour requirements, 249; mechanisation, 249; merchants, 114; physical requirements, 248; standard quantities, 25; transport, 115; versus other crops, 248
changes; acreage of crops, 245; agricultural land use, 145–8; agriculture, 323–5
Chatteris, 117
cheese; farmhouse, 130, 164–5, 170; in Scotland, 159
cherries, 291
chilling, 17, 18
Chisholm, M., 111, 133, 324
Church, B. J., 324
Cistercians, 211
Clapham, J. H., 17, 28
Clark, C., 330
Clarke, K. R. and K. E. Hunt, 10
climate, 30–41; averages, 30, 32; data, 30; effects, 39, 41; factors, 32; modification, 37; range, 32; variability, 32, 41
climatic requirements; conflicting, 40; critical, 40; for barley, 254; dairy cattle, 160; for fodder crops, 268; for glasshouses, 296, 298; for oats, 254; for orchards, 290; for pigs, 229; for potatoes, 265; for poultry, 236; for sheep, 210; for soft fruit, 295; for sugar beet, 262; for tomatoes, 299; for vegetables, 285; for wheat, 252
cloches; 41, 295
Clyde Valley, 170, 297, 299
cold storage, orchard crops, 111, 294
Coles, R., 244
combine harvesters, 101, 257; diffusion, 105

common land, England and Wales; acreage, 63; enumeration, 58, 62; farm size, 63; land improvement, 63; overstocking, 63, 209–10; sheep, 209–10;
common pasture, Scotland, 62, 70, 210
comparative advantage; between cereals, 256, 259; crops, 6, 247; enterprises, 303; orchards, 293; potatoes, 267; vegetables, 284
compass points, viii (note)
competition; cereals, 256, 257, 259; enterprises, 303; horticulture, 276
concentrates, 160
concentration; eggs; 242; horticulture, 301; orchards, 292, 294; pigs, 225, 233; poultry, 241
contracting; crops. 119, 288; eggs, 241; labour, 96; machinery, 68, 96, 99; pigs, 125; poultry, 244; sales, 112, 119, 125; soft fruit growing, 295; sugar beet, 122; work, 68
co-operatives; cereals, 112; crops, 112; eggs, 241; horticulture, 112; machinery, 68; marketing, 112, 288; orchard fruit, 294; pigs, 230; potatoes 112; poultry 244
Coppock J. T. 10, 79, 151, 324, 329
Corn Laws, 11, 14
Cornwall, 133, 285, 302
costs, *see also* transport costs; cultivation, 44, 47; heating glasshouses, 299; herd size, 161, 162; labour, fodder, 269; labour, horticulture, 277; labour, orchards, 294; labour, potatoes, 266; labour, soft fruit, 295; labour, sugar beet, 263; marketing glasshouse crops, 299
costs, shared, 5; beef production, 180; livestock markets, 126
Cotswolds, 82, 212
cottages, 74
'county' farm, 3
County War Agricultural Executive Committees, 23
Covent Garden, 116–9, 132
Cracknell, B. E., 86, 106, 273
Crimean War, 14
Crofters Act 1886, 70

Crofters' Commission, 70
crofting counties, 60, 70
crofts, 60; neglected, 70
crops, 245–73, *see also* under individual crops; acreage changes, 245; cereals, 248–60; choice of, 246; climate, 32; combinations, 318; environment, 30; flowers, 275, 280–9; fruit, 275, 289–96; fodder, 268; glasshouse, 275, 296–300; horticultural on farms, 274; horticulture, 274–302; hygiene, 246, 258; index of productivity, 50; markets, 113–23, 247; mechanisation, 249; physical requirements, 247, 248; processing, 282; quality, 247; restraints, 247; retained on farm, 245, 247; roots, 260–7; rotations, 246; sales, 245; successive, of barley, 246; tradition, 248; type of farm, 312, 320, 321; vegetables, outdoor, 275, 280–9; yields, 245, 247
crop residues, beef cattle, 192
crop rotations, 246; herbicides and pesticides, 246; leases, 246; plant hygiene, 246, 258; purpose, 246
Crowe, P., vii
cucumbers, 298
cultivation; limits of, 29, 43; variation in costs, 44, 47
Cupar, 123, 261–2

dairy cattle; artificial insemination, 157; breeds, 156–7; distribution, 153, 158–9; feed, 160, 161; herd replacements, 5, 158; herd size, 153, 175; labour requirements, 162; seasonal distribution, 158–9; surplus calves, 158; yields, 157
dairy farming, 152–78, *see also* milk production; climate, 160; definition, 176; enterprises, 304; farm buildings, 163; farm size, 161; grazing season, 161; herd size, 153, 175; increasing herd size, 325; land suitable, 161; markets, 163–4; other enterprises, 176, 178; produce, 14, 16, 20; small holdings, 177; specialisation, 173; type of farming, 173–8, 320, 321

Dairy Regulations, 75
Darlington, Peter, Ltd., 258
Davey, B. H., 330
Davidson, B. R., 215, 221
Davies, E. T., 79
Davies, William, 141
day degrees, 36
day length, 33
dealers, 124; store livestock, 109
decision making, 82
deer, 207, 211; rough grazing, 143–4
deficiency payments, wheat, 249, 258
dehydration, horticultural crops, 277
Denman, D. R., 71
Denmark, 225, 233
Department of Agriculture and Fisheries for Scotland, vi, viii (note), 10
depopulation, 12; uplands, 144
derationing, 20, 26–7
Devon, 111, 288
diet, 14, 113; horticultural crops, 14; pig meat, 224; poultry, 224; regional changes, 14; regional differences, 14, 113, 178, 218, 267; vegetables, 14; wheat, 14
diffusion, 105
disease; blight, 266; cattle plague, 165; cereal susceptibility, 258; climate, 40; control by spraying, 266; eelworm, 265; 'eyespot', 252; foot and mouth, 325; potatoes, 265; prediction, 266; 'take all', 252; virus (potatoes), 266; wheat, 252, 258
distilleries, 115
distribution, spatial, 3–4
Dixey, R. N., 77, 79
Donaldson, J. G. S. and P., 10
drainage, 31; impeded, 45; improved, 40; land quality, 45; natural, 45; poor, 45; under-, 45; war-time, 23; wheat needs, 252; yields, 45
ducks, 234
Duffield, B. S., 329
Dufftown, 115
Dundee, 172
Dunn, J. M., 221
Dunsford, W. J., 79

INDEX

Dutch; dairy produce, 20; horticulturalists, 297

earliness; climate, 46; potatoes, 265; strawberries, 295
ecology, applied, 29
economic geography, 9
economic motives, 4
economies of scale; cereals, 249; dairying, 162; horticulture, 288; orchards, 294; pigs, 232; poultry, 243
Edinburgh, 172
efficiency, 68
egg production, 234, 237–42
eggs, 17, 20, 21, 235, 237–42; marketing, 131–2, 241; packing stations, 131; producer-retailers, 131; supplies, 17; transport costs, 131–2
Ellison, W., 45
Elms, C. E., 57
embanking, 31
empire produce, 21
enclosure, parliamentary, 73; farmsteads, 74; fragmentation, 73
Engledow, Sir Frank, vii
enterprises, 303–8; agricultural census, 304; 305; choice of, 3–8, 66; combinations, 318; competition, 303; definition, 304; England and Wales, 305; identification, 305, 315; intensive, 314; large businesses, 322; livestock units, 306; Scotland, 306; standard man-days, 94, 306–7; starch equivalent, 306; type of enterprise area, 314, 317; type of farming, 304, 319
environment; physical, 29; artificial, 41; man-made, 6, 31; natural, 6
Eskdalemuir, 298
estates, 69, 70
European Economic Community, 25, 325
Evans, H. G., 221
Evesham, 117
extensive systems, 63; enterprises, 66; farms, 318; labour, 66, 95; pigs, 229; poultry, 234; sheep, 200

factories; bacon, 126, 230; jam, 295; milk, 127; sugar beet, 262

factory farms, 74
Farm Improvement Scheme, 26, 74, 77
farm buildings, 73–5, see also housing
farm layout, 72–9; accessibility, 72; fragmentation, 73; location of buildings, 73–4; quality of land, 54; regional variation, 54; significance, 77; types of land, 73
Farm Management Survey, 6, 67, 93, 96, 273
farm size, 55–68; cereals, 61; changes, 56–61; crops and grass, 58; dairying, 161–2; distribution, 58–63; efficiency, 67–8; enlargement, 55; England and Wales, 64; enterprises, 66; flexibility, 67; herd size, 153–6; intensiveness, 66; land quality, 60; pigs, 231; poultry, 236; Scotland, 65; sheep, 209; significance, 66–8; size of business, 63–5; total area, 60
farm worker, 87–93; age, 83–4; casual, 91; decline, 12, 95, 97; definition, 80; foremen, 82; full-time, 83; holdings with, 88; numbers, 12, 88; part-time, 91; seasonal hours, 95; specialisation, 83; status, 87, 88
farmer; attitudes, 4, 249, 259; changes, 97–8; definition, 80; employers, 86; farm type, 87; hobby, 87; migrant, 87; motives, 4; numbers, 84; occupations, 84–5; part-time, 84, 85, 87; progressive, 66; ratio to workers, 93; relatives, 92, 93, 162; retirement, 66; tenant farmer, 68–72; wives, 92
farming, type of, 309–23, see also each enterprise; agricultural census, 304; arable rearing and feeding, 314; areas, 303, 314, 317; beef cattle, 193–9; classification, 303, 309–10, 314; crofting, 60; cropping, 312, 314; crops by regions, 269; dairy cattle, 173–8; dairy farms, 313; degree of dominance, 312, 313; England and Wales classification, 310, 320; extensive, 318; farm crops and farm type, 269; flexibility, 67,

309; full-time holdings, 309; high farming, 145; hill sheep farms, 314; horticulture, 300–1, 312; intensive, 314, 318; large businesses, 322; linked holdings, 56; livestock, 312; miners' holdings, 60; mixed holdings, 310; multiple holdings, 56–7; parish unit, 316; part-time holdings, 314; pigs, 233; poultry, 237; predominantly dairying, 312; qualitative classification, 310; quantitative classification, 310; rearing with arable, 314; rearing with intensive livestock, 314; regional, 1, 319–23; Scotland, classification, 313, 319; sequence of classification, 313; sheep, 220–1; size of unit, 316; small holdings, 60; statistical rules, 310; subjective approach, 309; upland farms, 314

farms, 54–79; animal housing, see housing; boundaries, 55; boundary maps, 74; buildings, 55, 61, 73–5, 163; characteristics, 5; distribution, 58–63; enclosure, 73; regional variation, 58–63; shape, 72, 73

farms, large; change, 67; efficiency, 67, 68; enterprises, 66; mechanisation, 68, 103; progressive farmers, 61; tractor use, 68

farms, small; change, 67; dairying, 161, 162; efficiency, 67, 68; feedingstuffs, 66; intensive enterprises, 66; labour, 66, 68; mechanisation, 68, 103

feedingstuffs; barley, 253, 257; cattle, 257; cereals, intensive beef, 192; egg production, 239; horses, 256; imported, 16, 268; oats, 254, 256; pigs, 222, 224, 229, 257; poultry, 222, 236; purchased for dairy farms, 161; sugar beet, 261; surplus vegetables, 288; wheat, 249

Felsted, 263

Fenland; mediaeval, 31; carrot growing, 97; yields, 265

fertilisers, 49–50; effects on land, 47; rate of application, 49, 50; subsidy, 23

fertility, soil, 31, 47
fields, 74–8
Finance Act 1961, 26
fixed equipment, 54, 78
flexibility, 67, 309; decreasing, 143
floods, 45
flowers, 275, 280–9; acreage, 280; distribution, 280, 284; glasshouse, 298; output, 280
flying flocks, 217
F.M.C. Ltd., 111, 126, 230, 233
fodder crops, 268
Folley, R. R. W., 274, 299, 301, 302
food industries, 28
Food, Ministry of, 23, 24
foot and mouth epidemic, 325
forestry, 13, 213
fragmentation, 73
fridd, 144, 216
frost, 37; effects on vegetables, 37, 285; frost-free period, 37; orchards, 37, 291
fruit, 14, 19, 20, 21, 275, 289–96; distribution, 289; imported, 20, 276; orchard, 289–94; output, 289; soft fruit, 295–6
Fylde, 236

game, management, 72
Garner, F. H., 10
Gasson, R., 87, 134, 274, 328
geese, 234
geology, drift, 46
Giles, A. K., 106
Giles, R. A., 299, 302
Gillmor, D., 329
glasshouse crops, 296–300 and see individual crops; air pollution, 298; climate, 298; development, 297; Dutch immigrants, 297; flowers, 298; heating, 297; holdings, 300; labour, 298; marketing, 299; output, 296; temperature, 298; yields, 278
glasshouses, 33, 41, 279; automatic watering, 299; heated, 295; heating costs, 41; obsolete, 298; temperature, 299
Glasgow, 172
'golden hoof', 211
golf courses, 56
gooseberries, 295

INDEX

government intervention, 20-8
grain driers, 105, 114
grants, 23-6; improvement, 23; post-war, 25-6; 'relevant production grants', 25; silo, 141
grass; improved, 268; starch equivalent, 306
grassland, 139-43; climate, 39, 141; distribution, 140; grazing, 140; management for beef, 191-2; mowing, 141; silage, 141; survey, 141
grazing; livestock, 30, 140; rough grazing, 143-4, 207-8; season, 36, 140-1; soils, 47; units, 148
Grigg, D. B., 79
gross margin, 214
Gross National Product, 28
gross output, 67, 272
groups of farmers or producers, 112; broiler selling, 244; horticulture, 288; pigs, 230
grouse, 144, 207
growth, plant; elevation, 43; growing season, 34; onset, 34; rate, 34; soils, 47

hail, 39
Hardaker, J. B., 273
Hart, J. F., 221
harvesting, 34-41; dates, 34, 36
hay, 141; climate, 39; distant markets, 109; yields, 39
Hay, F. G., 273
hedgerows, removal, 143
Henshall, J. D., 327, 330
high farming, 145
hill cow subsidy, 27, 189
hill sheep farming, 314; inflexibility, 67, 309
hill sheep subsidy, 27, 208; reduced rate, 26
Hilton, N., 53
Hinton, W. L., 287, 300, 301, 302
Hirsch, G. P., 94
hirsel, 216
hobby farming, 85, 87; beef, 188
Hogg, W. H., 30
holdings, agricultural, 55-79, *see also* farms; amalgamation, 58; farm type, 87; full-time, 84-5; miners' part-time, 60; multiple, 56; non-viable, 66; part-time, 56, 84-5; size, 64; small, 60; spare-time, 56; squatters, 60
Holland, 50, 247, 284, 286
home farm, 69, 85, 86
Home Grown Cereals Authority, 25, 113
Hop Marketing Board, 22, 247, 268
hops, 21, 22, 267-8
horses, 27, 256; oats, 256; release of land, 103, 149
Horticultural Improvement Scheme, 26, 293, 297
horticultural markets; contract sales, 89; crops, 119-20; grading, 121; links, 117; outlets, 117-9; primary, 116; processing factories, 275, 277; transport, 119
horticulture, 21, 24-5, 274-302, *see also* under individual crops; acreage, 277; advantages of home-grown, 276; capital investment, 278; characteristics, 275; climate, 33; competitive, 276, 301; concentration, 285, 287, 300; crops under glass, 275, 296-300; definition, 274; development, 285, 301; distribution, 278, 279, 285; elasticity of demand, 275; farmer-growers, 287, 288, 300; flowers, 275, 280-9, 300; fruit, 275, 289-96, 300; government support, 278; imports, 20, 276; intensiveness, 277; labour requirements, 277, 288; land classification, 52; localisation, 321; market connections, 279; market gardens, 274, 300; natural protection, 276; optimum location, 301; orchards, 289-94; outdoor vegetables, 275, 280-9; out of season crops, 276; output sales, 274, 280, 301; quotas, 276; reputation, 279; specialised labour, 279, 289; speculative crops, 278; standard of living, 275; tariffs, 276; transport, 280; type of farm, 274, 300, 312; variety, 274; varying profits, 278; yields, 277
housing, livestock; broiler houses, 224, 234, 239, 243; dairy cattle, 163; hill cows, 188; hill sheep,

housing livestock—*contd.*
205–6; pigs, 229; poultry, 236;
urban cowsheds, 164
Howe, G. M., 53
Hughes, R. E., 43
Hull, 297
human consumption, *see also* diet;
barley, 254; oats, 254; potatoes,
264; sugar, 261; wheat, 249
Hunt, K. E. and K. R. Clarke, 10
Huntingdonshire, 76

imports, 14–22; agricultural produce, 276; bacon, 21, 225, barley, 253; changing, 14–7; competition, 16, 21, 22; dairy produce, 16, 17, 152; early potatoes, 263; feedingstuffs, 268; fruit, 20, 276; hops, 268; horticultural produce, 276; Irish store cattle, 181, 195, 197; Irish pullets, 238; maize, 16; meat, 21, 179, 201; seed potatoes, 266; vegetables, 20, 276; wheat, 12, 17, 249, 256
Ingersent, K. A., 323
inputs, 5
insurance, 40
intensive systems, 63; ducks, 234; enterprises, 66; farm size, 66–7; farms, 314, 318; feedingstuffs, 66; glasshouse crops, 296, 301; horticultural crops, 274, 277; labour, 66, 94–5; pigs, 223; pigs and poultry, 222; poultry, 234, 239, 243; turkeys, 234
interdependence, 4, 200
inter-farm differences, 9
investment, *see* capital investment
Ireland, 12; store cattle, 181, 195, 197; point-of-lay pullets, 238
irrigation, 38, 40; orchards, 291; potatoes, 266; vegetables, 285
Isle of Ely, 50
isopleths, 316
Italian horticulturalists, 20, 290

Jones, G. E., 105, 106

kale, 268
Kendall, M. G., 50
Kent, 85, 214

Kidderminster, 262
Kirk, J. H., 112, 134

labour, 80–4; casual, 91, 268; cereals, 249; changes, 97; contract, 96; costs, 162–3; dairy farms, 162; family, 84–92; farm size, 94–5; farm type, 96–7; female, 227; hops, casual, 268; horticulture, 277–8, 288–9; inputs, 96; orchards, 294; peaks, 95–6, 206, 263, 266, 294; picking beans and raspberries, 277; pigs, 232; potatoes, 226, 266, 267; poultry, 243; productivity and herd size, 161; requirements, seasonal, 95, 266, 267, 277; requirements, regional, 94; requirements, standard, 64, 307; roots 261, 263, 266–7; sheep, 206; soft fruit, 295; standard mandays, 94, 149, 306; sugar beet, 263; supply and demand, 5, 96; units, 93–4; vegetables, 289
laissez-faire, 7, 21
lamb; imports, 21, 201; production, 201; regional consumption, 113
Lanarkshire, 210
Lancashire, 237–8
land, *see also* agricultural; semi-agricultural, 56; uncultivable, 46
land classification, 50–3; acreage, 51; categories, 51; enterprises, 52; farm layout, 77; farm practices, 52; management, 51; mechanisation, 103; revaluation, 52
land improvement, 23, 31, 257
land ownership, 71
land reclamation, 23
land tenure, 68–72; changes, 69–70; land in hand, 70; mixed, 69; significance, 77; types, 68–9
land use; multiple, 144; other, 13, 28
land-use surveys, 30
Land Utilisation Survey, v
land values, 78–9
land, economic definition, 54
landlord/tenant, 71
Lands Tribunal, 72

INDEX

Langdon, A. L., 78
lapse rate, 33
Lea Valley, 279, 297–8
leaching, 38, 49
lease restrictions, 68
Leeds, 116
lettuce; glasshouse, 298; out door, 282
light, 33, 298
lime, 23, 31; barley, 259; subsidy, 23
limestone, 46
limiting conditions, 30–1
Lindsey, 284–5, 286–7
Littleport, 279
Liverpool, 110, 238
livestock, *see also* under individual types; climate, 32; contribution, 19–20, 148; distribution, 148; leading, 149; man-days, 149; movements of beef cattle, 180, 190; of calves, sample study, 189; of dairy cattle, 158; of pigs, 228; of sheep, 204, 217–8; products, 18–9; shelter, 40–1; trends, 149
livestock changes, 149; composition, 151; increased stocking, 148; regional distribution, 151
livestock marketing, 123–7; auctions, 124; buyers, 125; dead-weight sales, 125; dealers, 124; fatstock, 124–6; marketing groups, 124–5; pigs, contract sales, 125; role, 126; slaughter-houses, 126; store, 124; transport costs, 126
livestock units, 148, 150; enterprise data, 305; enterprises, 305, 306; type of farm, 312; type of farming, 320–1
living, standard of, 4; 1850–1965, 13; regional differences, 13
localisation, 3, 29; choice vegetables, 284; cucumbers, 298; glasshouses, 298; hops, 267; horticulture, 278, 288; orchards, 289, 292; persistence, 279; raspberries, 295; soft fruit, 295; sugar beet, 262
locational advantage, 18; dairy farming, 173; effects of milk marketing schemes, 168–9; theory, 163
Locke, G. M. L., 75

Long, W. H., 180, 199

McEwan, L. W., 124, 134
McIntosh, F., 199
McQueen, J. D. W., 130, 159, 178
machinery, *see* mechanisation
Mackney, D., 53
maize, 16, 29
malting barley, 253, 254, 256
management, 9; differences, 29
managers, 82
mangolds, 268
Manley, G. M., 43
manufacture, *see* processing
manufacturing industry, interdependence, 27
manure, 27, 192, 211, 279
man-days, *see* standard man-days
maps, 2, 3, *see* list of figures, p. xii. *See also* Ordnance Survey; boundaries, 55; farms located, 55; geological, 30; grassland, 142; National Atlas Series, 50; topographical, 30; tithe commutation, 55; type of farming, 315
March, 115
market gardens, *see* horticulture
marketing, 107–34, *see also* under cereals, horticulture, livestock, milk, potatoes; buyers, 108, 125, 130; characteristics, 108; contract sales, 112, 119, 125; data on cereals, 107; farm gate sales, 109, 131, 134; farmers' groups, 112, 124, 125, 230, 244, 288; links, 112, 115, 117; middlemen, 108, 114; perishability, 111; processors, 109, 115, 119, 122, 131; producer-retailers, 109, 131; transport, costs, 109, q.v.; wholesalers, 109
marketing boards, *see* under individual items; revived, 24–5
markets, 107–34; cereals, 107, 113–6; crops, 247; eggs, 239; glasshouse crops, 299; horticultural produce, 116–21; livestock, 123–7; milk, 127–31, 164–73; orchard fruit, 294; pigs, 230–1; poultry, 236; roots, 121–3; sheep, 218; soft fruit, 295
marling, 31
Martin, V., 329

Maunder, A. H., 77, 79
meat, 14, 16, 17, 18; poultry meat, 235, 242–4
Meat and Livestock Commission, 25, 126, 179, 233
mechanisation, 27, 98–106; cereal crops, 249, 257; contract, 99; data, 98; farm layout, 102; farm and field enlargement, 61, 77, 103; hops, 268; horticulture, 277; index of, 101; input, 98; labour savings, 104; machinery syndicates, 106–7; potatoes, 266; roots, 261; soft fruit, 295; sugar beet, 263; terrain, 44, 102; transfers, 99; trends and effects, 103–6; type of farming, 102
meteorological data, 30
Meteorological Office, 32; classification for milk production, 160
Mid-Wales Report, 84
military training grounds, 72
milk, 16–22, 127–31; consumption, 177–8; evaporated, 19; liquid, 18; long-keeping, 17
milk marketing, 127–31; boards, 22, 128, 134, 166–8, 176; characteristics, 127; depots, 129, 130, 165, 167; dependence on fresh milk market, 152; factories, location, 127, 167–9, 172; factory products, 130; farmhouse manufacture, 164–73; prices, 127–8, 166–72; sales, 152; seasonal changes, 130; tankers, 130, 169; transport, 128–30
Milk Marketing Board, the, 128, 134, 176
milk marketing boards, 22, 128, 166; creation, 166; effects, 168; marketing schemes, 166
milk prices, 127–8, 166–72; collapse, 166; interregional compensation, 167; London, 166; maximum retail, 127; pool, 167; producer, 169–70, 172
milk producers; decline, 169–70; registered, 152
milk production; importance, 152; manufacture, pigs, 230; markets, 164–73; natural protection, 152; rise, 152; seasonal, 159; transport, 164–6; trends, England and Wales, 164–70; trends, Scotland, 170–3
Millman, R., 79
Ministry of Agriculture, vi (note), viii (note), 10
Ministry of Defence, 72
mixed corn, 248
moisture, 37–41
Montgomeryshire, 62
moorland, 31; moorland edge, 43, 143
motives, farmers', 4; re beef cattle, 180, 199
motor vehicles, 18
Munton, R. J. C., 10, 53, 328, 329
Murray, K. A. H., 28
mutton, 21, *see also* lamb; imports, 21, 201; production, 201

Napolitan, L., 324
National Atlas, 50
National Farm Survey, 55, 68, 73, 77, 78
National Farmers' Union, collective bargaining, 166
national fertiliser survey, 75
National Grid, type of farming classification, 316
National Institute of Agricultural Botany, 36
National Milk Investigations, 161
Natural Wheat Survey, 6, 45, 253
National Resources (Technical) Committee, 221, 231
Newcastle, 166
New Forest, 62, 143
New Zealand, imported meat, 16–19, 179, 201
Norfolk, 284–5, 286–7
Norfolk 4-course, 211
Norris, J. M., 53
North America, 14–15
North Atlantic Drift, 33
Northumberland, 199
Nottinghamshire, 73

oats, 22, 254; declining demand, 256; distribution, 254, 257; fodder, 254; human consumption, 254; physical requirements, 254; plant breeders, 257; regional importance, 257; winter keep scheme, 256; yields, 254

occupiers, 68–72, 84–7; corporate, 86; owner-occupiers, 68–9; partners, 86; status and significance, 86; tenants, 68–72
Ojala, E., 19
optimum location, 30
orchards, 289–94; arable depression, 292; cider, 292; climate, 290; cold stores, 294; commercial, 289; concentration, 292, 294; development, 292; distribution, 289; frost, 37, 291; grubbing up, 292, 293; irrigation, 291; labour, 294; localisation, 289; marketing, 294; non-commercial, 290; packing stations, 294; planting grants, 293; plums, 292; sites, 291; size, 290; specialisation, 294; yields, 290, 293, 294
Ordnance Survey, *see also* maps; farm-boundary, 55; grassland, 142; National Atlas Series, 50; type of farming, 315
Orkney, 129, 242
Ottawa Agreements 1932, 21
owner-occupiers, 68–9
Oxford Farming Conference, 325
Oxfordshire, 77

packing stations; egg, 131–2; apples, 294
parish data, 3
pears; dessert, 291; perry, 292
peas, 268; canning, 283; harvesting dry, 282, 283; quick-freezing, 282, 283, 286
peat, 47; fen, 48
Penzance, 110
permanent grass, 139, 141–8; definition, 139; distribution, 141; factors affecting, 141; quality, 141–2; Scottish agricultural census, 139; unploughable, 139; wartime ploughing, 23
Philips, R. W., 43
physical factors, 29–53
physiographic factors, 41–9
Pig Industry Development Authority (PIDA), 228, 229, 233
pigs, 222–34; breeding, 228; climate, 229; consumption, 224; cycle, 223; distribution, 225–8; distribution, factors affecting, 228, 234; enterprise, 305; extensive systems, 229; family labour, 232; farm size, 231; fattening, 228; feed, 222, 229–30; fluctuations, 222; herd size, 225–7, 231–2; holdings, 231; housing, 41, 229; intensive systems, 222, 229; labour, 232; land requirement, 222; marketing, 125, 233; movements, 228; numbers, 224; organisation, 223; output, 222; products, 224
pigs and poultry; common characteristics, 222–4; land requirements, 222; organisational change, 223–4; rapid fluctuations, 222–3
planning control, 26
ploughing grants, 26, 146
ploughing, one-way, 44
plums, 292
poaching (treading), 140; dairy cattle, 161; sheep, 214
policy, agricultural, 11–28; consequences, 23, 28
population, 11–4; farming, 87–93; growth, 11; rural, 12; urbanisation, 11
pork, consumption, 224
potato blight, 40, 266
potatoes, 263–7; acreage per grower, 265; canning, 267; climate, 265; comparative advantage, 267; crisps, 267; dehydration, 267; demand inelastic, 263; distribution, 264–6; earlies, 30, 264, 265; enterprises, 304; fluctuating yields, 263; holdings with, 264; human consumption, 243, 267; imports of earlies, 263; imports of seed potatoes, 266; irrigation, 266; main crop, 264; marketing, 121–2; physical requirements, 265; processing, 122, 267; quality, 265; registered growers, 121, 265; risk with earlies, 265; seed potatoes, 264, 266; specialisation, 265; spraying, 266; variety, 122; yield and physical conditions, 265

Potato Marketing Board, 22, 121–2, 247, 264; basic acreages, 121; licensed merchants, 122; registered growers, 121
potential evaporation, 38
poultry, 14, 222–4, 234–44; breeding, 234; broilers, 224, 234, 243–4; consumption, 224; distribution, 235, 236; egg production, 237–42; enterprises, 305; extensive, 234; feed, 222, 236; flock size, 240; fluctuations, 223; housing, 41, 236, 239–40, 243; intensive, 234, 239, 243; labour, 241, 243; land requirements, 222; marketing, 131–2, 243; meat, 224, 242–3; numbers, 224, 239; organisation, 223; output, 222
price, movements of, 4
price review, annual, 24, 259
prices, guaranteed, 22, 24, 25; bacon, 22; barley, 259; eggs, 239; milk, 22, 25, 166; oats, 254, 259; pigs, 233; sheep, 22; sugar beet, 25; wheat, 259
processing, 27; canning, 14, 177, 275, 286, 295; chilling, 17, 18; dehydration, 122, 277; distilling, 115; malting, 235, 254, 256; milling, 109, 230; quick-freezing, 119, 275, 277, 286, 295; refrigeration, 14, 16, 17; sugar beet pulping, 122; of cereals, 114–5; of eggs, 131; of fruit, 295; of horticultural crops, 277; of milk, 130; of pigs, 230; of potatoes, 122; of sugar beet, 122; of vegetables, 282
Proctor (barley), 258
producer-retailers, 109; eggs, 131; milk, 130, 166–8, 172
protection, natural, 16, 18–19, 152
proximity, urban, 18, 20, *see also* marketing

quality, 26, 121, 265
quick-freezing, 111, 119; horticultural crops, 275, 277, 286; soft fruit, 295
quotas, 20; hop acreage, 22; horticultural crops, 276, 278; potato acreage, 121; voluntary, 25

railways, 14–5, 17–8; grain, 115; Great Britain, 16–7; horticulture, 119, 279; marketing, 110; milk, 164–5; North America, 14–5; sugar beet, 123
rain-days, 37
rainfall, 27–39
rape, 268
raspberries, 295
rationing, food, 24
reaper binders, 102, 105, 257
refrigeration, 14, 16, 17
regional accounts, 8
regional diets, 218, 230–1
regionalisation; dairying, 146, 173; increasing, 9; land use, 14, 151; livestock, 151, 173; pigs, 228, 233; poultry, 241, 243
regions, agricultural, 8
Reid, I., 47
relief, 43–6
rent; average county, 78; cereal farming rise, 249; farm type, 79; land quality, 78; open market rent, 27
Reorganisation Committee for Eggs, 131, 244
road transport; effects, 18; grain, 115; horticultural crops, 119, 280, 286; marketing, 110; sugar beet, 122–3
Robertson, I. M. L., 106
Romney Marsh, 142, 148, 201, 208, 210, 218
roots, 260–7; characteristics, 260; cleaning crops, 260; labour requirements, 261; marketing, 121–3; mechanisation, 261; potatoes, 263–7; sugar beet, 261–3; transport costs, 122–3, 260; yields, 260
rough grazing, 143–4, 207–8; character, 207–8; contribution, 143; distribution, 143; factors affecting, 144; feed values, 144; improvement, 143; multiple uses, 144; past uses, 143
Rowntree, B. S., 28
Runciman Report, 107, 116, 117, 119, 120, 133, 134
rye, 248

sales; crops 245; dairy produce, 152; fat cattle, 179; hops, 267; horticulture, 274; pigs and poultry, 222; roots, 260; sheep, 200
Salt, J., 329
Sanders, H. G., 27
Sandwich, 285
Sandy, 285
savoys, 282
Scola, P. M., 106, 324
Scottish Dairy Census, 1965, 153
Scottish Land Court, 72
Seafield, Countess of, 71
seasonality; eggs, 242, 293; labour, 95; labour for sheep, 206; labour in orchards, 294; mechanisation, 95; milk marketing, 130; milk production, 159; pigs, 228; sheep, 218–9; turkeys, 234
Second Agricultural Revolution, 245
Second World War; agricultural achievements, 24; agricultural policy, 23
Self, P., 28
share-cropping, poultry, 239
sheep, 200–21; breeds, 204–5, 216–7; capital equipment, 205–6; common grazings, 209–10; complementary role, 214; cross-breeding, 216–7; development in uplands, 211; distribution, 201–10; distribution, changing, 211–4; enterprises, 305; fairs, 124, 218; farm size, 209; farm type, 219–21; feeding, 207–8; flock composition, 213; flock size, 206–7; folding, 211–2; holdings with, 202–4; housing, 205–6; labour requirements, 206–7; lambing dates, 210; lambing rates, 219; lowlands, 211–2, 217; mortality, 210; movements, 144, 204, 217–8; manurial value, 211; output, 200, 220; place in economy, 214; products, 201; ratio to other livestock, 202; role in agricultural geography, 200; seasonality, 218–9; stratification, 214–7; transhumance, 144, 211; uplands, 205–6; utilisation of poor pasture, 210
shelter belts, 40

shepherds, availability, 206; club-stocks, 210
Shetland, 133
shipping, 15, 17
Short, B. M., 329
silage, 141
Simpson, E. S., 178
slope, 44; livestock, 44; machinery, 44
small-holdings, 70–1; committee of enquiry, 71; statutory, 70
Smith, L. P., 39, 53
Smith, Wilfred, 8
snow, 39
Snowdonia, 43
soft fruit, 295–6; cloches, 295; contract growing, 295; declining acreage, 296; distribution, 295; grown on farms, 295; labour requirements, 295; transport costs, 295; yields, 296
spatial distribution, 3
soils, 46–9; acid brown, 47; acidity, 38, 254, 257; agricultural, 46; barley, 254; brown earths, 47; calcareous, 47; cereals, 248; clay with flints, 284; climate, 47; dairying, 161; fertilisers, 49–53; fertility, 31, 46–9; first class (A) land, 284; grey-brown podsolic, 47; heavy, 47; hops, 268; limestone, 46; liming, 257; moisture deficit, 38; oats, 254; orchards, 291; parent materials, 46; pigs, 229; potatoes, 265; roots, 261; shallow, 50; significance, 47; stony, 50; successive cropping, 246; sugar beet, 262; vegetables, 284; wheat, 252
Somerset, 111
specialisation; crops, 246; glasshouse crops, 299; orchards, 294; potatoes, 265; raspberries, 295
specialist holdings; pigs, 228, 233; poultry, 237, 242
Spencer, John, 8
squatters, 60
Stamp, L. D., 10, 50, 53, 81
standard labour requirements, 64; enterprises, 307
standard man-days (s.m.ds.), 94, 150; enterprises, 306; livestock, 149; stability, 306

standard outputs, enterprises, 306
standard quantities, 25; milk, 128, 169, 172
Stapledon, Sir George, 141
starch equivalent, 136
statistical data, 3
stock, *see* livestock
storage, cold, for top fruit, 111, 294; grain, by merchants, 114; grain, on farms, 114
Storing, H. J., 28
stratification, sheep farming, 214–7
strawberries; Blackpool disaster, 297; early, 295
subsidies, 20; barley acreage, 258; beef cow, 188; calf, 183; hill cow, 27, 189; hill sheep, 26, 27, 208; oats acreage, 258; postwar, 25; sugar beet, 20; wartime, 23; wheat acreage, 258; winter keep, 27, 208
Suffolk, 286
sugar, 17, 20, 21; Commonwealth producers, 261; home-grown, 20, 261; human consumption, 261; imports, 11, 17, 20, 21
sugar beet, 261–3; acreage limited, 261; British Sugar Corporation, 113, 122, 248, 261; closure of Cupar factory, 261; distribution, 261; factories, 122, 262; fodder, 261; holdings, 262; labour costs, 263; marketing, 122–3; mechanisation, 263; soils, 262; subsidy, 20; suitable land, 261; transport policy, 122, 261; varieties, monogerm, 263; vegetable growing, 286, yields, 245, 262
sunshine, 33
surpluses, agricultural, 24–5
swedes, 268
Sykes, J. D., 273

tariffs; European, 16; general, 20; horticultural crops, 276, 278
Tarrant, J., 316, 329
Taylor, J. A., 30, 53
temperature, 34–42; accumulated, 34
temporary grass, 139–42; break crop, 143; distribution, 142; grass under 7 years old, 139;
government encouragement, 139; ploughing grants, 139; quality, 142
tenant farmers, 68–72; compensation, 68; distribution, 69; notice to quit, 72; restrictions, 71; Scotland, 69; security, 70, 72
theory, vii
Thomas, E., 57
thunderstorms, 39
tillage, 135–9; decline, 147; distribution, 136–8; favourable factors, 136; fertilisers, 138; harvesting, 138; mechanisation, 138; output, 135; rainfall, 138
Tinline, R., 329
tithe maps, 55
Tomato and Cucumber Marketing Board, 24–5
tomatoes, 298
topography, effects, 46
tractors; crawler tractors, 44; distribution, 101; horsepower, 98; tasks, 101; trends, 104
trade and transport, 14–20
traditional practices, 4; beef, 180; crops, 248; sheep, 214
transhumance, sheep, 144, 211
transport, 14–20; bulky commodities, 111; cart, 12; development of horticulture, 285–6; diesel lorry, 286; flexibility of road transport, 18; grain, 115; horticultural produce, 119, 280; improvements, 17; international, 18; milk, 127–30, 164–73; motor lorry, 286; rail, 110, 285; rationalisation, 111, 168–9; return loads, 111; road, 18, 110, 286; speed, 111; store stock, 124 sugar beet, 122
transport costs, 17; difficult to assess, 109; eggs, 131; grain, 115; horticultural produce, 133; livestock, 110–1, 126; marketing, 132, 133; milk, 129–33, 164–8; new potatoes, 133; potatoes, 267; roots, 260; soft fruit, 295; store sheep, 133; sugar beet, 122; tapering rates, 172; uniform rates, 18, 230
trend surfaces, 316
trespassing, 18

INDEX

turkeys, 234
turnips, 268

United States of America, 17
uplands, contribution of sheep, 215
urbanisation, 11–4

veal, 158, 182
vegetables, outdoor, 280–9; acreage, 280; advantage of home-grown, 19; characteristics, 275; choice produce, 284; climate, 285; coarse, 284; diet, 14; distribution, 280, 284; holdings, 282; imports, 20, 276; irrigation, 285; markets, 288; output, 280; processing, 282; requirements, 275, 283; stock feed, 288; tariffs, 21; variability of supply, 108
vegetation; deterioration, 213; uplands, 207–8
Verdon-Smith Report, 107, 124, 126, 133, 134
Vince, S. W. E., 81
voluntary agreements, 20
Von Thünen, 107, 108, 164, 327

wages, disparity, 80
Warley, T. K., 134
warping, 31
water, livestock requirements, 39
water table, 45
weather, *see also* climate; effect on horticultural crops, 278
Weaver, John, 2; Crop Combinations, 318
weeds, 258; in root crops, 261; in wheat crops, 258; pre-emergence herbicides, 277
Welland, 190
West, W. J., 44
West Riding, 111, 287
wethers, decline, 213

wheat, 249–53; cash crop, 249; costs, 253; deficiency payments, 249, 258; diseases, 252; distribution, 252, 256, 258; extent during Second World War, 31; feeding stuffs, 249; guaranteed prices, 25, 259; human consumption, 14, 249; imports, 12, 17, 249, 256; in rotation, 258; mechanisation, 249; physical requirements, 252; restricted outlets, 258; risks, 253; standard quantities, 25; uses, 249; yield, 252
Wheat Act 1932, 21, 22, 256
Wibberley, G. P., vi, 10, 215, 221
Wigtownshire, 166, 172
Williams, H. T., 10
Williams, W. M., 106
wind, 39
winter keep scheme, 27, 208; fodder crops, 269; oats, 254
wintering; ewe hoggs, 144; other sheep, 216
Wise, M. J., 329
Wolpert, J., 330
wool, contribution, 200; imports, 200; marketing board, 24; prices, 19
World Agricultural Census 1960, 69
Worthing, 298

Yetminster, 73
yields, cereals, 256–7; crops, 14, 245, 247; effects of weather, 278; eggs, 240; fluctuations, 247; hay, 39; high on small acreages, 31; horticulture, 277; land classification, 50; milk, 157; orchards, 290, 293, 294; potatoes, 265, 266; roots, 260; soft fruit, 296; sugar beet, 245, 262; varying, in horticulture, 278; wheat, 14, 252, 253